# Geometry

## Applying • Reasoning • Measuring

# Chapter 8
# Resource Book

The Resource Book contains the wide variety
of blackline masters available for Chapter 8.
The blacklines are organized by lesson. Included
are support materials for the teacher as well as
practice, activities, applications, and assessment
resources.

 McDougal Littell
A HOUGHTON MIFFLIN COMPANY
Evanston, Illinois • Boston • Dallas

## Contributing Authors

The authors wish to thank the following
individuals for their contributions to the
Chapter 8 Resource Book.

*Eric J. Amendola*
*Karen Collins*
*Michael Downey*
*Patrick M. Kelly*
*Edward H. Kuhar*
*Lynn Lafferty*
*Dr. Frank Marzano*
*Wayne Nirode*
*Dr. Charles Redmond*
*Paul Ruland*

ISBN: 0-618-02071-3

123456789-VEI- 04 03 02 01 00

# Contents

**8  *Similarity***

# Contents

# Contents

## Descriptions of Resources

This Chapter Resource Book is organized by lessons within the chapter in order to make your planning easier. The following materials are provided:

**Tips for New Teachers** These teaching notes provide both new and experienced teachers with useful teaching tips for each lesson, including tips about common errors and inclusion.

**Parent Guide for Student Success** This guide helps parents contribute to student success by providing an overview of the chapter along with questions and activities for parents and students to work on together.

**Prerequisite Skills Review** Worked-out examples are provided to review the prerequisite skills highlighted on the Study Guide page at the beginning of the chapter. Additional practice is included with each worked-out example.

**Strategies for Reading Mathematics** The first page teaches reading strategies to be applied to the current chapter and to later chapters. The second page is a visual glossary of key vocabulary.

**Lesson Plans and Lesson Plans for Block Scheduling** This planning template helps teachers select the materials they will use to teach each lesson from among the variety of materials available for the lesson. The block-scheduling version provides additional information about pacing.

**Warm-Up Exercises and Daily Homework Quiz** The warm-ups cover prerequisite skills that help prepare students for a given lesson. The quiz assesses students on the content of the previous lesson. (Transparencies also available)

**Activity Support Masters** These blackline masters make it easier for students to record their work on selected activities in the Student Edition.

**Alternative Lesson Openers** An engaging alternative for starting each lesson is provided from among these four types: *Application, Activity, Geometry Software,* or *Visual Approach.* (Color transparencies also available)

**Technology Activities with Keystrokes** Keystrokes for Geometry software and calculators are provided for each Technology Activity in the Student Edition, along with alternative Technology Activities to begin selected lessons.

**Practice A, B, and C** These exercises offer additional practice for the material in each lesson, including application problems. There are three levels of practice for each lesson: A (basic), B (average), and C (advanced).

# Contents

**Reteaching with Additional Practice** These two pages provide additional instruction, worked-out examples, and practice exercises covering the key concepts and vocabulary in each lesson.

**Quick Catch-Up for Absent Students** This handy form makes it easy for teachers to let students who have been absent know what to do for homework and which activities or examples were covered in class.

**Cooperative Learning Activities** These enrichment activities apply the math taught in the lesson in an interesting way that lends itself to group work.

**Interdisciplinary Applications/Real-Life Applications** Students apply the mathematics covered in each lesson to solve an interesting interdisciplinary or real-life problem.

**Math and History Applications** This worksheet expands upon the Math and History feature in the Student Edition.

**Challenge: Skills and Applications** Teachers can use these exercises to enrich or extend each lesson.

**Quizzes** The quizzes can be used to assess student progress on two or three lessons.

**Chapter Review Games and Activities** This worksheet offers fun practice at the end of the chapter and provides an alternative way to review the chapter content in preparation for the Chapter Test.

**Chapter Tests A, B, and C** These are tests that cover the most important skills taught in the chapter. There are three levels of test: A (basic), B (average), and C (advanced).

**SAT/ACT Chapter Test** This test also covers the most important skills taught in the chapter, but questions are in multiple-choice and quantitative-comparison format. (See *Alternative Assessment* for multi-step problems.)

**Alternative Assessment with Rubrics and Math Journal** A journal exercise has students write about the mathematics in the chapter. A multi-step problem has students apply a variety of skills from the chapter and explain their reasoning. Solutions and a 4-point rubric are included.

**Project with Rubric** The project allows students to delve more deeply into a problem that applies the mathematics of the chapter. Teacher's notes and a 4-point rubric are included.

**Cumulative Review** These practice pages help students maintain skills from the current chapter and preceding chapters.

## LESSON 8.1

**COMMON ERROR** Students tend to write ratios without indicating the units of measure. This usually results in their failure to convert unlike units to like units before simplifying the ratio. Emphasize the correct solution format illustrated in Example 1 on page 457.

**TEACHING TIP** Consider using pieces of string and rulers if tape measures are not available for investigating the Activity on page 457. Allow students to compare ratios with others seated near them or collect and display the data for all to see. Discuss similarities and differences as time allows.

## LESSON 8.2

**INCLUSION** Visual models are an asset to students with limited English proficiency. Consider bringing some examples from home to illustrate ratios, such as various sizes of canned goods, cracker or cereal boxes, and books. Examining similar models in this lesson and the next can be very beneficial. Students also could bring in small model toys for discussion.

## LESSON 8.3

**COMMON ERROR** The introduction of the symbol for similarity is in this lesson. Students may confuse the following symbols: $\cong$, $\approx$, and $\sim$. Caution them to be careful when writing them.

**TEACHING TIP** The term *scale factor* is important in this lesson. Students may have already used it earlier in this chapter. They probably have encountered scale factors in other classes such as Science or History when reading maps. Ask them to identify other times when they have encountered the use of scale factors.

**TEACHING TIP** Theorem 8.1 on page 475 addresses the fact that for similar polygons the ratio of the perimeters is equal to the ratio of corresponding sides. A question may arise concerning the ratio of the areas of similar figures. For a time efficient discussion exploring that idea, consider using squares or right triangles with side

lengths that are Pythagorean triples (e.g., 3-4-5, 6-8-10, 9-12-15). List the ratios of corresponding sides, the ratio of the perimeters, and the ratio of the areas. It should be clear that the ratio of the areas is the square of the ratio of corresponding side lengths.

## LESSON 8.4

**TEACHING TIP** A discussion of the Activity on page 480 leads nicely into Postulate 25 on page 481. Students should be able to conclude that when two pairs of angles are congruent, the third pair of angles in the triangles is also congruent. Their reasoning might come from the use of the Triangle Sum Theorem on page 196, or from the Third Angles Theorem on page 203. They should realize that with the Angle-Angle Similarity Postulate, the assumption of having similar triangles is accurate and more direct when using two pairs of congruent angles.

**TEACHING TIP** Stress the general concept about similar polygons given before Example 5 on page 482. Students should realize that the ratio of any two corresponding lengths is equal to the scale factor of two similar polygons. Such length measures include sides, perimeters, altitudes, medians, angle bisector segments and diagonals, but not area, as area is not a measurement of length. Stress that this concept applies to all polygons, not just triangles.

**COMMON ERROR** Students may find diagrams confusing when triangles are drawn in an overlapping manner as in Example 5 on page 482. Advise them to first write the ratio or proportion using letter labels and check that they have the correct corresponding parts identified. Then they can substitute measures for the appropriate segments. Sometimes it is helpful to pull the diagram apart and draw the triangles separately.

## LESSON 8.5

**TEACHING TIP** For demonstrations, check to see if a pantograph is available somewhere in an art class, an architecture class, or even somewhere in your school. You could also make one or have a

## *Tips for New Teachers*

**For use with Chapter 8**

student make one like the model in Example 4 on page 490. Use cardboard strips, paper fasteners, and a suction cup. It is a good idea to experiment using it before demonstrating it in the classroom.

**TEACHING TIP** Stress again the importance of writing the proportions first and then substituting measurements for the lengths of segments as in Examples 3, 4, 5, and 6 on pages 489–491. Students should always check to make sure they have identified the correct pair of corresponding parts in their proportions or ratios. This practice should be continued throughout the rest of the chapter.

## LESSON 8.6

**TEACHING TIP** Use the Activity on page 500 to discuss the use of congruent angles to achieve parallel lines. Explore and identify relationships in part 4 of the construction. A segment has been divided into four equal parts, but what about the triangles that are formed in the process? Students should recognize that the angles are corresponding angles.

## LESSON 8.7

**TEACHING TIP** The properties of a dilation indicate that the scale factor, $k$, should not equal 1. Be

prepared for students to ask what would happen if $k$ did equal 1. Any point $P$ would map onto itself when $k = 1$. The resulting dilation image would not be a reduction or an enlargement; it would simply be the same as the preimage. You should be aware that a dilation with a scale factor of 1 is an example of an *identity transformation*. Identity transformations are not covered in this course, but students may encounter them in future studies of mathematics.

**TEACHING TIP** Remind students that any ratio is a fraction that can be written as a decimal and then as a percent. In Example 3 on page 508, the problem asks students to express how much larger the shadow is than the puppet as a percent. In part (a), be aware that the original proportion uses

$$\frac{CP}{SH}.$$

Since the shadow is an enlargement of the puppet the scale factor must use

$$\frac{SH}{CP}.$$

You may interpret the result of 1.25 as 125%. 100% accounts for the original puppet and the additional 25% the amount of increase.

---

## *Outside Resources*

### BOOKS/PERIODICALS

Shilgalis, Thomas W. "Finding Buried Treasures—An Application of the Geometer's Sketchpad." *Mathematics Teacher* (February 1998); pp. 162–165.

### ACTIVITIES/MANIPULATIVES

Reinstein, David, Paul Sally, and Dane R. Camp. "Generating Fractals through Self-Replication." *Mathematics Teacher* (January 1997); pp. 34–38, 43–45.

### SOFTWARE

Bellemain, Franck and Jean-Marie Laborde. *Cabri Geometry II*. Dallas, TX; Texas Instruments, 1995.

### VIDEOS

Apostol, Tom M. *Similarity*. With program guide/workbook. Reston, VA; NCTM.

---

NAME _____ DATE _____

# Parent Guide for Student Success

**For use with Chapter 8**

**Chapter Overview** One way that you can help your student succeed in Chapter 8 is by discussing the lesson goals in the chart below. When a lesson is completed, ask your student to interpret the lesson goals for you and to explain how the mathematics of the lesson relates to one of the key applications listed in the chart.

| Lesson Title | Lesson Goals | Key Applications |
|---|---|---|
| **8.1: Ratio and Proportion** | Find and simplify the ratio of two numbers. Use proportions to solve real-life problems. | • Painting<br>• Planets' Gravity<br>• Gulliver's Travels |
| **8.2: Problem Solving in Geometry with Proportions** | Use properties of proportions. Use proportions to solve real-life problems. | • Titanic Model<br>• Blueprints<br>• Lolo Trail |
| **8.3: Similar Polygons** | Identify similar polygons. Use similar polygons to solve real-life problems. | • Poster Design<br>• TV Screens<br>• Total Eclipse |
| **8.4: Similar Triangles** | Identify similar triangles. Use similar triangles in real-life problems. | • Tourmaline Crystal<br>• Aerial Photography<br>• The Great Pyramid |
| **8.5: Proving Triangles are Similar** | Use similarity theorems to prove that two triangles are similar. Use similar triangles to solve real-life problems. | • Scale Drawings<br>• Rock Climbing<br>• Unisphere |
| **8.6: Proportions and Similar Triangles** | Use proportionality theorems to calculate segment lengths. Use proportionality theorems to solve real-life problems. | • Insulating an Attic Room<br>• Lot Prices<br>• New York City Map |
| **8.7: Dilations** | Identify dilations. Use properties of dilations to create real-life perspective drawings. | • Shadow Puppets<br>• Enlarging Photos<br>• Perspective Drawing |

## Study Strategy

**Connect to the Real World** is the study strategy featured in Chapter 8 (see page 456). Have your student make a list of the main topics in the chapter. Then, work together to give a real-world example for each topic. Your student may remember how to work problems related to the topic by remembering the real-world example.

NAME _____ DATE _____

# Parent Guide for Student Success

For use with Chapter 8

**Key Ideas** Your student can demonstrate understanding of key concepts by working through the following exercises with you.

| Lesson | Exercise |
|--------|----------|
| 8.1 | Find the ratio of 2.5 feet to 1 yard in simplest form. |
| 8.2 | Company A manufactures a jet that has 410 seats and costs $6567 an hour to operate. Company B manufactures a jet that costs $4885 an hour to operate. How many seats should the jet have in order for it to have the same rate per seat as the Company A jet? |
| 8.3 | A formal, rectangular garden is 20 meters wide and 25 meters long. You want to make a similar garden in your yard that is 8 meters wide. How much edging do you need to enclose your garden? |
| 8.4 | A right triangle with a 15-centimeter hypotenuse has a 50° angle. Another right triangle has a 10-centimeter hypotenuse and a 50° angle. How do you know the triangles are similar? |
| 8.5 | A cliff casts a 36-foot shadow at the same time you cast a 2-foot shadow. You are standing level with the bottom of the cliff. If you are 5 feet tall, what is the height of the cliff? |
| 8.6 | You have a piece of property with the back line perpendicular to the sides. The front of the lot along the road is 132 meters long and on a slant. You divide your lot into three lots to sell, by making lines parallel to the original sides. The backs of the new lots are 35 m, 42 m, and 33 m wide. The 35-meter lot has 42 meters of road frontage. Find the road frontage of each of the other two lots. |
| 8.7 | Find the coordinates of the vertices of a dilation of the triangle with vertices $A(-2, 0)$, $B(1, 3)$, and $C(2, -1)$, with the origin as the center and a scale factor of 2. |

## Home Involvement Activity

**You Will Need:** A yardstick, cardboard, scissors, tape
**Directions:** Find the overall dimensions of the outside of your home. You may need to use indirect measurement. Make a scale model of your home out of cardboard. First choose a scale factor. Then make a table with the actual dimensions and the dimensions of your scale model. Use the table to construct the actual model.

**8.6:** 50.4 m and 39.6 m   **8.7:** $A'(-4, 0)$, $B'(2, 6)$, $C'(4, -2)$

**8.1:** $\frac{5}{6}$   **8.2:** about 305 seats   **8.3:** 36 m   **8.4:** Angle-Angle Similarity Postulate   **8.5:** 90 ft

**Answers**

NAME _____ DATE _____

# Prerequisite Skills Review

For use before Chapter 8

**EXAMPLE 1** *Finding Perimeter*

Find the perimeter of the given figure.

a.
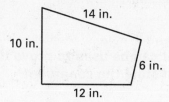
14 in.
10 in.
6 in.
12 in.

b.
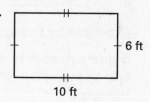
6 ft
10 ft

**SOLUTION**

a. perimeter = 10 + 12 + 6 + 14

= 42 in.

b. perimeter = 6 + 10 + 6 + 10

= 32 ft

## Exercises for Example 1

Find the perimeter of the given figure.

1.
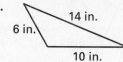
14 in.
6 in.
10 in.

2.
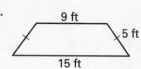
9 ft
5 ft
15 ft

3.
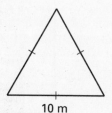
10 m

4.
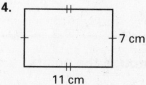
7 cm
11 cm

5.
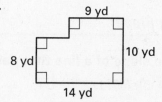
9 yd
8 yd
10 yd
14 yd

6.
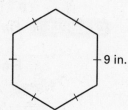
9 in.

**EXAMPLE 2** *Using Postulates or Theorems to State Triangles are Congruent*

Name a postulate or theorem that can be used to state the two triangles are congruent. Then state the congruent triangles.

a.
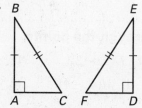
B
E
A    C  F    D

b.
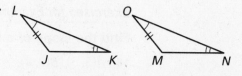
L
O
J       K   M       N

# Prerequisite Skills Review

**For use before Chapter 8**

## SOLUTION

**a.** HL Congruence Theorem, $\triangle ABC \cong \triangle DEF$

**b.** AAS Congruence Theorem, $\triangle JKL \cong \triangle MNO$

### Exercises for Example 2

Name a postulate or theorem that can be used to prove the two triangles are congruent. Then state the congruent triangles.

**7.**

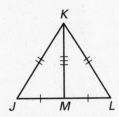

**8.**

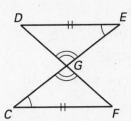

**9.**

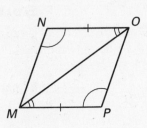

**10.**

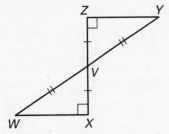

**11.**

**12.**

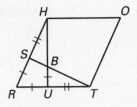

---

**EXAMPLE 3** *Finding Slope*

Find the slope of a line that passes through the points.

**a.** $A(-4, 3), B(7, 1)$          **b.** $A(0, 5), B(-1, 4)$

### SOLUTION

**a.** $m = \dfrac{3-1}{-4-7}$

$= -\dfrac{2}{11}$

**b.** $m = \dfrac{5-4}{0-(-1)}$

$= \dfrac{1}{1} = 1$

### Exercises for Example 3

Find the slope of a line that passes through the points.

**13.** $A(5, 3), B(8, 4)$

**14.** $C(-6, 1), D(2, -2)$

**15.** $E(0, 4), F(5, 2)$

**16.** $G(6, -3), H(3, -3)$

**17.** $L(-1, -4), M(4, -1)$

**18.** $J(-4, 9), K(-4, 2)$

**Geometry**
Chapter 8 Resource Book

# Strategies for Reading Mathematics

**For use with Chapter 8**

## Strategy: Using Diagrams

When you are solving a problem like the one in the example below that involves similar figures, you should draw a diagram. Be sure to carefully label the diagram so that corresponding sides are easy to find.

### EXAMPLE

You have a 6 inch by 9 inch rectangular photo that you want to reduce. You want the reduction to be 2 inches wide. How long will it be?

### SOLUTION

Draw a diagram to show the original photo and its reduction. Use $x$ to label the length of the reduction.

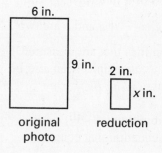

lengths of photos →→ $\dfrac{x \text{ in.}}{9 \text{ in.}} = \dfrac{2 \text{ in.}}{6 \text{ in.}}$ ←← widths of photos

$$x = \frac{2}{6} \cdot 9$$

$$x = 3$$

The length of the reduction will be 3 inches.

---

### STUDY TIP

**Drawing a Diagram**

Be sure to use the same orientation when drawing similar figures. For example, in the figures above, both figures are drawn with the lengths in a vertical orientation and the widths in a horizontal orientation.

---

## Questions

1. Suppose you drew the original photo in the above example as shown at the right. Show how you would draw the reduction in your diagram.

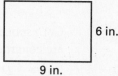

2. Suppose you want the reduction to be 3 inches wide. How would you change the diagram above to help you solve the problem? Use the new diagram to find the length.

3. Suppose you want to enlarge the original photo so that the enlargement will be 12 inches long. Draw a new diagram that you can use to help you find the width of the enlargement. Then find the width.

4. Suppose you want to enlarge the original photo so that the enlargement will be 12 inches wide. How would you change the diagram you drew in Question 3 to help you solve the problem? Use the new diagram to find the length.

## Visual Glossary

The Study Guide on page 456 lists the vocabulary for Chapter 8 as well as review vocabulary from previous chapters. Use the page references on page 456 or the Glossary in the textbook to review key terms from prior chapters. Use the visual glossary below to help you understand some of the key vocabulary in Chapter 8. You may want to copy these diagrams into your notebook and refer to them as you complete the chapter.

### GLOSSARY

**ratio of *a* to *b*** (p. 457) The quotient $\frac{a}{b}$ if *a* and *b* are two quantities that are measured in the same units. Can also be written as *a* : *b*.

**proportion** (p. 459) An equation that equates two ratios.

**means of a proportion** (p. 459) The middle terms of a proportion.

**extremes of a proportion** (p. 459) The first and last terms of a proportion.

**similar polygons** (p. 473) Two polygons such that their corresponding angles are congruent and the lengths of corresponding sides are proportional.

### Solving Proportions

Suppose the ratio $\frac{x}{4}$ is equal to the ratio $\frac{9}{12}$. Then the folowing proportion can be written.

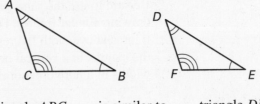

means ⟶     extremes

$$4(9) = 12x$$

The product of the means equals the product of the extremes.

$$36 = 12x$$

$$3 = x$$

### Corresponding Parts of Similar Polygons

Similar polygons have corresponding parts. The order of the vertices in a similarity statement shows the corresponding vertices.

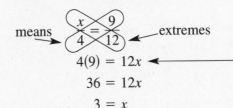

Triangle *ABC*    is similar to    triangle *DEF*.

$$\downarrow \qquad \downarrow \qquad \downarrow$$

$$\triangle ABC \qquad \sim \qquad \triangle DEF$$

Corresponding angles are congruent.

$$\angle A \cong \angle D$$
$$\angle B \cong \angle E$$
$$\angle C \cong \angle F$$

Corresponding sides are proportional.

$$\frac{AB}{DE} = \frac{BC}{EF} = \frac{CA}{FD}$$

TEACHER'S NAME _____ CLASS _____ ROOM _____ DATE _____

# *Lesson Plan*

**1-day lesson** (See *Pacing the Chapter,* TE pages 454C–454D)          **For use with pages 457–464**

**GOALS**
1. **Find and simplify the ratio of two numbers.**
2. **Use proportions to solve real-life problems.**

State/Local Objectives _____

_____

✓ **Check the items you wish to use for this lesson.**

## STARTING OPTIONS

____ Prerequisite Skills Review: CRB pages 5–6
____ Strategies for Reading Mathematics: CRB pages 7–8
____ Homework Check: TE page 440: Answer Transparencies
____ Warm-Up or Daily Homework Quiz: TE pages 457 and 444, CRB page 11, or Transparencies

## TEACHING OPTIONS

____ Motivating the Lesson: TE page 458
____ Lesson Opener (Application): CRB page 12 or Transparencies
____ Examples 1–7: SE pages 457–460
____ Extra Examples: TE pages 458–460 or Transparencies
____ Closure Question: TE page 460
____ Guided Practice Exercises: SE page 461

## APPLY/HOMEWORK
### Homework Assignment

____ Basic    12, 16, 20, 24, 28, 32, 36, 40, 44, 45–47, 54–58, 62–66, 68–76 even
____ Average    12, 16, 20, 24, 28, 32, 36, 40, 44, 45–47, 51–58, 62–66, 68–76 even
____ Advanced    12, 16, 20, 24, 28, 32, 36, 40, 44, 45–47, 51–58, 62–76 even

### Reteaching the Lesson

____ Practice Masters: CRB pages 13–15 (Level A, Level B, Level C)
____ Reteaching with Practice: CRB pages 16–17 or Practice Workbook with Examples
____ Personal Student Tutor

### Extending the Lesson

____ Applications (Real-Life): CRB page 19
____ Challenge: SE page 464; CRB page 20 or Internet

## ASSESSMENT OPTIONS

____ Checkpoint Exercises: TE pages 458–460 or Transparencies
____ Daily Homework Quiz (8.1): TE page 464, CRB page 23, or Transparencies
____ Standardized Test Practice: SE page 464; TE page 464; STP Workbook; Transparencies

Notes _____

_____

_____

TEACHER'S NAME _____ CLASS _____ ROOM _____ DATE _____

# *Lesson Plan for Block Scheduling*

Half-day lesson (See *Pacing the Chapter,* TE pages 454C–454D)          For use with pages 457–464

**GOALS**   1. **Find and simplify the ratio of two numbers.**
             2. **Use proportions to solve real-life problems.**

State/Local Objectives _____

_____

_____

✓ **Check the items you wish to use for this lesson.**

**STARTING OPTIONS**
_____ Prerequisite Skills Review: CRB pages 5–6
_____ Strategies for Reading Mathematics: CRB pages 7–8
_____ Homework Check: TE page 440: Answer Transparencies
_____ Warm-Up or Daily Homework Quiz: TE pages 457 and
         444, CRB page 11, or Transparencies

**TEACHING OPTIONS**
_____ Motivating the Lesson: TE page 458
_____ Lesson Opener (Application): CRB page 12 or Transparencies
_____ Examples 1–7: SE pages 457–460
_____ Extra Examples: TE pages 458–460 or Transparencies
_____ Closure Question: TE page 460
_____ Guided Practice Exercises: SE page 461

**APPLY/HOMEWORK**
**Homework Assignment**
_____ Block Schedule: 12, 16, 20, 24, 28, 32, 36, 40, 44, 45–47, 51–58, 62–66, 68–76 even

**Reteaching the Lesson**
_____ Practice Masters: CRB pages 13–15 (Level A, Level B, Level C)
_____ Reteaching with Practice: CRB pages 16–17 or Practice Workbook with Examples
_____ Personal Student Tutor

**Extending the Lesson**
_____ Applications (Real-Life): CRB page 19
_____ Challenge: SE page 464; CRB page 20 or Internet

**ASSESSMENT OPTIONS**
_____ Checkpoint Exercises: TE pages 458–460 or Transparencies
_____ Daily Homework Quiz (8.1): TE page 464, CRB page 23, or Transparencies
_____ Standardized Test Practice: SE page 464; TE page 464; STP Workbook; Transparencies

| CHAPTER PACING GUIDE | |
| --- | --- |
| **Day** | **Lesson** |
| 1 | Assess Ch. 7: **8.1 (all)** |
| 2 | 8.2 (all); 8.3 (begin) |
| 3 | 8.3 (end); 8.4 (begin) |
| 4 | 8.4 (end); 8.5 (begin) |
| 5 | 8.5 (end); 8.6 (begin) |
| 6 | 8.6 (end); 8.7 (begin) |
| 7 | 8.7 (end); Review Ch. 8 |
| 8 | Assess Ch. 8; 9.1 (all) |

Notes _____

_____

_____

Lesson 8.1

NAME ———————————————————— DATE ————

# WARM-UP EXERCISES

For use before Lesson 8.1, pages 457–464

## Simplify each fraction.

**1.** $\dfrac{12}{15}$

**2.** $\dfrac{14}{56}$

**3.** $\dfrac{21}{6}$

**4.** $\dfrac{90}{450}$

················································································

# DAILY HOMEWORK QUIZ

For use after Lesson 7.6, pages 437–444

**Describe the transformations that will map the frieze
pattern onto itself.**

**1.** ∝/∝/∝/∝/∝/∝/∝/∝

**2.** ⊃⊂⊃⊂⊃⊂⊃⊂⊃⊂⊃⊂

**3.** Use the design below to create a frieze pattern with the
classification THG.

## *Application Lesson Opener*

For use with pages 457–464

A useful way to compare two quantities is to write a *ratio*. If your family has 2 dogs and 3 cats, the ratio of dogs to cats is $\frac{2}{3}$, or 2:3. If you also have 4 fish, the ratio of dogs to cats to fish is 2:3:4.

**Write the indicated ratio for the members in your class today. Then create some ratios of your own about your class.**

1. boys : girls

2. girls : boys

3. students : teachers

4. left-handed : right-handed

5. freshmen : sophomores

6. 14-year-olds : 15-year-olds : 16-year olds

7. not currently employed : currently employed

8. have no pet at home : have at least one pet at home

9. born in this state : not born in this state

10. ate cereal for breakfast : did not eat cereal for breakfast

11. not wearing jeans today : wearing jeans today

12. birthday is in spring : birthday is in summer :
    birthday is in fall : birthday is in winter

**The girls' soccer team won 10 games and lost 2, and the boys' soccer team won 12 games and lost 3.**

1. What is the ratio of the girls' wins to their losses?

2. What is the ratio of the boys' wins to their losses?

3. What is the ratio of the girls' wins to the total number of games played?

4. What is the ratio of the boys' wins to the total number of games played?

5. Which team had the greater winning ratio?

**Simplify the ratio.**

6. $\dfrac{6 \text{ yards}}{12 \text{ yards}}$

7. $\dfrac{14 \text{ trucks}}{7 \text{ trucks}}$

8. $\dfrac{16 \text{ people}}{24 \text{ people}}$

9. $\dfrac{32 \text{ meters}}{24 \text{ meters}}$

**Find the width to length ratio of each rectangle. Then simplify the ratio.**

10.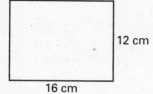

12 cm

16 cm

11.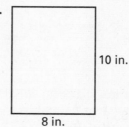

10 in.

8 in.

12.

8 in.

2 ft

**Rewrite the fraction so that the numerator and denominator have the same units. Then simplify.**

13. $\dfrac{2 \text{ yd}}{24 \text{ in.}}$

14. $\dfrac{60 \text{ mm}}{1 \text{ cm}}$

15. $\dfrac{40 \text{ g}}{1 \text{ kg}}$

16. $\dfrac{20 \text{ ft}}{3 \text{ yd}}$

17. $\dfrac{3 \text{ lb}}{12 \text{ oz}}$

18. $\dfrac{5 \text{ weeks}}{30 \text{ days}}$

19. $\dfrac{85 \text{ cm}}{0.5 \text{ m}}$

20. $\dfrac{2 \text{ mi}}{60 \text{ ft}}$

**Solve the proportion.**

21. $\dfrac{x}{3} = \dfrac{10}{15}$

22. $\dfrac{y}{10} = \dfrac{2}{5}$

23. $\dfrac{20}{30} = \dfrac{m}{120}$

24. $\dfrac{4}{x+2} = \dfrac{16}{x+5}$

25. $\dfrac{3}{y-2} = \dfrac{15}{y}$

26. $\dfrac{2}{y-3} = \dfrac{3}{y}$

27. On an N-gauge model train set, a tank car is 3.75 inches long. An actual tank car is 50 feet long. What is the ratio of the length of the actual car to the length of the model tank car?

**Simplify the ratio.**

1. $\dfrac{8 \text{ books}}{24 \text{ books}}$

2. $\dfrac{24 \text{ trees}}{14 \text{ trees}}$

3. $\dfrac{18 \text{ balls}}{36 \text{ balls}}$

4. $\dfrac{48 \text{ feet}}{36 \text{ feet}}$

**Rewrite the fraction so that the numerator and denominator have the same units. Then simplify.**

5. $\dfrac{2 \text{ qt}}{4 \text{ gal}}$

6. $\dfrac{250 \text{ mg}}{10 \text{ g}}$

7. $\dfrac{24 \text{ oz}}{2 \text{ lb}}$

8. $\dfrac{14 \text{ ft}}{6 \text{ yd}}$

9. $\dfrac{4 \text{ft}}{8 \text{ in.}}$

10. $\dfrac{4 \text{ days}}{36 \text{ hours}}$

11. $\dfrac{1.5 \text{ m}}{80 \text{ cm}}$

12. $\dfrac{440 \text{ yd}}{2 \text{ mi}}$

**Use the number line to find the ratio of the distances.**

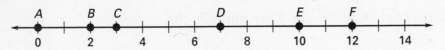

13. $\dfrac{AB}{CD} = $ ?

14. $\dfrac{BC}{DE} = $ ?

15. $\dfrac{AC}{BD} = $ ?

16. $\dfrac{CF}{AB} = $ ?

**Solve the proportion.**

17. $\dfrac{x}{6} = \dfrac{9}{24}$

18. $\dfrac{y}{9} = \dfrac{4}{6}$

19. $\dfrac{17}{24} = \dfrac{m}{120}$

20. $\dfrac{6}{x} = \dfrac{8}{x+3}$

21. $\dfrac{4}{y+3} = \dfrac{3}{y-4}$

22. $\dfrac{5}{2y-7} = \dfrac{3}{y}$

**The ratio of two side lengths of the triangle is given. Solve for the variable.**

23. $AB : BC$ is 2:5

24. $MN : MO$ is 3:4

25. $DE : EF$ is 8:5

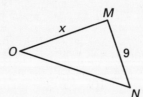

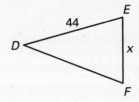

**In Exercises 26 and 27, use the following information.**

The largest submarines in the United States Navy are of the Ohio class. Each submarine is 560 feet long.

26. You purchase a scale model of one of the submarines. The package states the scale of 1 inch : 16 feet. What is the length of the completed model?

27. If the model is approximately 5 inches tall, what is the height of the actual submarine?

NAME _____ DATE _____

## *Practice C*

For use with pages 457–464

**Rewrite the fraction so that the numerator and denominator have the same units. Then simplify.**

1. $\dfrac{4 \text{ days}}{16 \text{ hours}}$

2. $\dfrac{18 \text{ yd}}{6 \text{ ft}}$

3. $\dfrac{0.6 \text{ km}}{200 \text{ m}}$

4. $\dfrac{80 \text{ mm}}{0.6 \text{ cm}}$

5. $\dfrac{5 \text{ gal}}{4 \text{ qt}}$

6. $\dfrac{18 \text{ in.}}{3 \text{ yd}}$

7. $\dfrac{25 \text{ min}}{2 \text{ hr}}$

8. $\dfrac{220 \text{ yd}}{3 \text{ mi}}$

**Solve the proportion.**

9. $\dfrac{x}{8} = \dfrac{5}{20}$

10. $\dfrac{y}{8} = \dfrac{6}{15}$

11. $\dfrac{5}{13} = \dfrac{m}{52}$

12. $\dfrac{8}{x} = \dfrac{12}{x+6}$

13. $\dfrac{6}{y+4} = \dfrac{5}{y-7}$

14. $\dfrac{5}{2y-7} = \dfrac{3}{y-3}$

**Solve.**

15. The perimeter of a rectangle is 40 feet. The ratio of the width to the length is 2:3. Find the length and the width.

16. The area of a rectangle is 192 square feet. The ratio of the width to the length is 3:4. Find the length and the width.

17. The measures of the angles in a triangle are in the extended ratio of 3:4:5. Find the measures of the angles.

18. The measures of the angles in a triangle are in the extended ratio of 2:5:8. Find the measures of the angles.

**You are given an extended ratio that compares the lengths of the sides of the triangle. Find the length of each side.**

19. $AC : BC : AB$ is 2:1:2

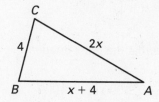

20. $AB : BC : AC$ is 3:4:2

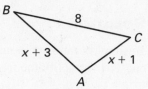

**In Exercises 21 and 22, use the following information.**

An architect wishes to represent the length of a room with a 5 inch segment.

21. If the room is actually 18 feet long, write the ratio of the scale length to the actual length.

22. Find the scale width, if the room is 12 feet wide.

NAME _____ DATE _____

# *Reteaching with Practice*

**For use with pages 457–464**

**GOAL** **Find and simplify the ratio of two numbers**

Lesson 8.1

---

### VOCABULARY

If $a$ and $b$ are two quantities that are measured in the same units, then the **ratio of $a$ to $b$** is $\dfrac{a}{b}$.

An equation that equates two ratios is a **proportion.**

In the proportion $\dfrac{a}{b} = \dfrac{c}{d}$, the numbers $a$ and $d$ are the **extremes** of the proportion and the numbers $b$ and $c$ are the **means** of the proportion.

**Properties of Proportions**

1. **Cross Product Property**   The product of the extremes equals the product of the means.

$$\text{If } \frac{a}{b} = \frac{c}{d}, \text{ then } ad = bc.$$

2. **Reciprocal Property**   If two ratios are equal, then their reciprocals are also equal.

$$\text{If } \frac{a}{b} = \frac{c}{d}, \text{ then } \frac{b}{a} = \frac{d}{c}.$$

---

**EXAMPLE 1** *Simplifying Ratios*

Simplify the ratios.

a. $\dfrac{8 \text{ in.}}{2 \text{ ft}}$

b. $\dfrac{1 \text{ km}}{500 \text{ m}}$

**SOLUTION**

To simplify ratios with unlike units, convert to like units so that the units divide out. Then simplify the fraction, if possible.

a. $\dfrac{8 \text{ in.}}{2 \text{ ft}} = \dfrac{8 \text{ in.}}{2 \cdot 12 \text{ in.}} = \dfrac{8}{24} = \dfrac{1}{3}$

b. $\dfrac{1 \text{ km}}{500 \text{ m}} = \dfrac{1 \cdot 1000 \text{ m}}{500 \text{ m}} = \dfrac{1000}{500} = \dfrac{2}{1}$

**Exercises for Example 1**

**Simplify the ratio.**

1. $\dfrac{25 \text{ cm}}{2 \text{ m}}$

2. $\dfrac{18 \text{ ft}}{2 \text{ yd}}$

3. $\dfrac{2 \text{ ft}}{24 \text{ in.}}$

4. $\dfrac{6 \text{ km}}{9 \text{ km}}$

---

**Geometry**
Chapter 8 Resource Book

# Reteaching with Practice

For use with pages 457–464

## EXAMPLE 2 Using Ratios

Triangle *XYZ* has an area of 25 square inches.
The ratio of the base of $\triangle XYZ$ to the height of
$\triangle XYZ$ is 2:1. Find the base and height of $\triangle XYZ$.

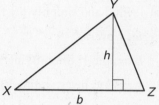

### SOLUTION

Because the ratio of the base to the height is 2:1, you can represent the
base as 2*h*.

$A = \dfrac{1}{2}bh$        Formula for the area of a triangle

$25 = \dfrac{1}{2}(2h)h$        Substitute for *A* and *b*.

$25 = h^2$        Simplify.

$5 = h$        Find the positive square root.

So, $\triangle XYZ$ has a base of 10 inches and a height of 5 inches.

### Exercise for Example 2

**5.** The area of a rectangle is 125 ft². The ratio of the width to the length is 1:5.
Find the length and the width.

## EXAMPLE 3 Solving Proportions

Solve the proportion.

$\dfrac{x}{8} = \dfrac{5}{4}$

### SOLUTION

$\dfrac{x}{8} = \dfrac{5}{4}$        Write original proportion.

$4x = 40$        Cross product property

$x = 10$        Divide each side by 4.

### Exercises for Example 3

**Find the value of *x* by solving the proportion.**

**6.** $\dfrac{9}{x} = \dfrac{2}{7}$      **7.** $\dfrac{5}{3} = \dfrac{5x}{6}$      **8.** $\dfrac{4}{x-4} = \dfrac{3}{x}$      **9.** $\dfrac{3}{x} = \dfrac{x}{12}$

NAME _____ DATE _____

# *Quick Catch-Up for Absent Students*

**For use with pages 457–464**

The items checked below were covered in class on (date missed) _____

**Lesson 8.1: Ratio and Proportion**

____ **Goal 1**: Find and simplify the ratio of two numbers. (pp. 457–458)

*Material Covered:*

    ____ Example 1: Simplifying Ratios

    ____ Activity: Investigating Ratios

    ____ Student Help: Look Back

    ____ Example 2: Using Ratios

    ____ Example 3: Using Extended Ratios

    ____ Student Help: Look Back

    ____ Example 4: Using Ratios

*Vocabulary:*

    ratio of *a* to *b*, p. 457

____ **Goal 2**: Use proportions to solve real-life problems. (pp. 459–460)

*Material Covered:*

    ____ Student Help: Skills Review

    ____ Example 5: Solving Proportions

    ____ Example 6: Solving a Proportion

    ____ Example 7: Solving a Proportion

*Vocabulary:*

    proportion, p. 459                extremes, p. 459
    means, p. 459

____ Other (specify) _____

_____

**Homework and Additional Learning Support**

    ____ Textbook (specify) pp. 461–464 _____

_____

    ____ *Reteaching with Practice* worksheet (specify exercises)_____

    ____ *Personal Student Tutor* for Lesson 8.1

NAME _____ DATE _____

# Real-Life Application: When Will I Ever Use This?

**For use with pages 457–464**

## Cooking

Many recipes found in cookbooks and magazines are often written to serve 4–6 people. If you have a particularly small or large family, work for a catering service or restaurant, or host a party, you may find it necessary to cut or multiply a recipe. It is important to proportionally increase or decrease the ingredients to maintain the same flavor. This is especially true when working with spices, for a small amount of a spice can drastically affect the taste.

### In Exercises 1–5, use the following information.

You are hosting a party at your house. You want to make enough pizza for 30 people. You find the following recipe for a medium pizza that serves 3 people.

**Ingredients**

| | |
|---|---|
| Flour | 3 cups |
| Salt | 1 Tablespoon |
| Yeast | 1 teaspoon |
| Sugar | $\frac{1}{8}$ cup |
| Water | 1 cup |
| Olive oil | 4 Tablespoons |
| Tomato | 10 ounces |
| Basil | 2 Tablespoons |
| Oregano | 4 Tablespoons |
| Garlic powder | 4 Tablespoons |

When you made one pizza, you found that you could make the crust larger by changing the flour amount to $4\frac{1}{2}$ cups. This made a large pizza that would serve 6 people. You decided to change all of the amounts proportionally so that you would only have to make five large pizzas.

1. What is the ratio of medium pizza ingredients to large pizza ingredients?

2. Find the amount of each ingredient for a large pizza.

3. You see that you only have 1 cup of garlic powder. Is this enough to make 5 pizzas? (16 Tablespoons = 1 cup)

4. How many pizzas can you make without going to the store for garlic?

5. Suppose you change the recipe so that you spread the garlic over 5 pizzas. Changing all the other ingredients proportionally, what would be the new amount of flour?

# Challenge: Skills and Applications

For use with pages 457–464

A *conversion factor* is a ratio in which the numerator and denominator are equivalent measurements, expressed in different units. For example:

$\dfrac{12 \text{ in.}}{1 \text{ ft}}$ and $\dfrac{1 \text{ ft}}{12 \text{ in.}}$ are conversion factors because 12 in. = 1 ft.

$\dfrac{1 \text{ kg}}{1000 \text{ g}}$ and $\dfrac{1000 \text{ g}}{1 \text{ kg}}$ are conversion factors because 1 kg = 1000 g.

**In Exercises 1–6, use the given information to write two conversion factors.**

1. 1 mi = 5280 ft

2. 1 ton = 2000 lb

3. 10 mm = 1 cm

4. 1000 m = 1 km

5. 1 lb = 16 oz

6. 1 km ≈ 0.621 mi

**In Exercises 7–14, multiply by an appropriate conversion factor to convert the quantity to the given units.**

When multiplying by a conversion factor, units that appear in both a numerator and a denominator can be divided out in the same manner as variables are divided out.

**Example:** 42 in., to feet: $42 \text{ in.} = 42 \text{ in.} \cdot \dfrac{1 \text{ ft}}{12 \text{ in.}} = 3.5 \text{ ft}$

7. 3.7 km, to m

8. 135 lb, to tons

9. 567 mm, to cm

10. 324 oz, to lb

11. 126 in., to ft

12. 300 km, to mi

13. 12.7 mi, to ft

14. 18 days, to seconds

15. The distance from Memphis, Tennessee, to Louisville, Kentucky, is about 320 miles. Convert this distance to kilometers.

16. The diameter of Venus at the equator is about 12,100 kilometers. Convert this distance to miles.

17. The rotation period of Saturn is about $10\frac{2}{3}$ hours. Convert this length of time to minutes.

18. The height of Mount Everest is about 29,000 feet. Convert this height to miles.

19. The Statue of Liberty weighs 225 tons. Convert this weight to ounces.

**Geometry**
Chapter 8 Resource Book

## LESSON 8.2

TEACHER'S NAME _____ CLASS _____ ROOM _____ DATE _____

# Lesson Plan

1-day lesson (See *Pacing the Chapter,* TE pages 454C–454D)  For use with pages 465–471

**GOALS**
1. **Use properties of proportions.**
2. **Use proportions to solve real-life problems.**

State/Local Objectives _____

_____

## ✓ Check the items you wish to use for this lesson.

### STARTING OPTIONS
____ Homework Check: TE page 461: Answer Transparencies
____ Warm-Up or Daily Homework Quiz: TE pages 465 and 464, CRB page 23, or Transparencies

### TEACHING OPTIONS
____ Motivating the Lesson: TE page 466
____ Lesson Opener (Calculator): CRB page 24 or Transparencies
____ Examples 1–4: SE pages 465–467
____ Extra Examples: TE pages 466–467 or Transparencies; Internet
____ Closure Question: TE page 467
____ Guided Practice Exercises: SE page 468

### APPLY/HOMEWORK
**Homework Assignment**
____ Basic   10–28 even, 29, 30, 33, 34, 38–40, 44, 45, 47–57
____ Average   10–28 even, 29–31, 33, 34, 38–40, 44, 45, 47–57
____ Advanced   10–28 even, 29–31, 33, 34, 36–40, 44, 45, 47–57

**Reteaching the Lesson**
____ Practice Masters: CRB pages 25–27 (Level A, Level B, Level C)
____ Reteaching with Practice: CRB pages 28–29 or Practice Workbook with Examples
____ Personal Student Tutor

**Extending the Lesson**
____ Applications (Interdisciplinary): CRB page 31
____ Challenge: SE page 471; CRB page 32 or Internet

### ASSESSMENT OPTIONS
____ Checkpoint Exercises: TE pages 466–467 or Transparencies
____ Daily Homework Quiz (8.2): TE page 471, CRB page 35, or Transparencies
____ Standardized Test Practice: SE page 471; TE page 471; STP Workbook; Transparencies

Notes _____

_____

_____

*Lesson 8.2*

TEACHER'S NAME _____ CLASS _____ ROOM _____ DATE _____

# *Lesson Plan for Block Scheduling*

Half-day lesson (See *Pacing the Chapter,* TE pages 454C–454D)          **For use with pages 465–471**

**GOALS**  1. **Use properties of proportions.**
2. **Use proportions to solve real-life problems.**

State/Local Objectives _____

_____

_____

| CHAPTER PACING GUIDE | |
|---|---|
| Day | Lesson |
| 1 | Assess Ch. 7: 8.1 (all) |
| 2 | **8.2 (all)**; 8.3 (begin) |
| 3 | 8.3 (end); 8.4 (begin) |
| 4 | 8.4 (end); 8.5 (begin) |
| 5 | 8.5 (end); 8.6 (begin) |
| 6 | 8.6 (end); 8.7 (begin) |
| 7 | 8.7 (end); Review Ch. 8 |
| 8 | Assess Ch. 8; 9.1 (all) |

✓ **Check the items you wish to use for this lesson.**

## STARTING OPTIONS

____ Homework Check: TE page 461: Answer Transparencies
____ Warm-Up or Daily Homework Quiz: TE pages 465 and
        464, CRB page 23, or Transparencies

## TEACHING OPTIONS

____ Motivating the Lesson: TE page 466
____ Lesson Opener (Calculator): CRB page 24 or Transparencies
____ Examples 1–4: SE pages 465–467
____ Extra Examples: TE pages 466–467 or Transparencies; Internet
____ Closure Question: TE page 467
____ Guided Practice Exercises: SE page 468

## APPLY/HOMEWORK

**Homework Assignment  (See also the assignment for Lesson 8.3.)**
____ Block Schedule:  10–28 even, 29–31, 33, 34, 38–40, 44, 45, 47–57

## Reteaching the Lesson

____ Practice Masters: CRB pages 25–27 (Level A, Level B, Level C)
____ Reteaching with Practice: CRB pages 28–29 or Practice Workbook with Examples
____ Personal Student Tutor

## Extending the Lesson

____ Applications (Interdisciplinary): CRB page 31
____ Challenge: SE page 471; CRB page 32 or Internet

## ASSESSMENT OPTIONS

____ Checkpoint Exercises: TE pages 466–467 or Transparencies
____ Daily Homework Quiz (8.2): TE page 471, CRB page 35, or Transparencies
____ Standardized Test Practice: SE page 471; TE page 471; STP Workbook; Transparencies

Notes _____

_____

_____

NAME ________________ DATE ________

Available as a transparency

# WARM-UP EXERCISES

For use before Lesson 8.2, pages 465–471

**Solve each proportion.**

1. $\frac{x}{4} = \frac{5}{8}$

2. $\frac{10}{9} = \frac{25}{x}$

3. $\frac{2}{y+1} = \frac{6}{y}$

4. $\frac{x-2}{5} = \frac{x+4}{35}$

# DAILY HOMEWORK QUIZ

For use after Lesson 8.1, pages 457–464

1. The perimeter of a living room is 64 ft. The ratio of width to length is 3:5. What are the dimensions of the living room?

2. The measures of the angles of a triangle are in the extended ratio of 1:3:5. Find the measures of the angles.

**Solve the proportion.**

3. $\frac{a}{8} = \frac{6}{20}$

4. $\frac{10}{x+6} = \frac{2}{x}$

LESSON 8.2

NAME ______________________ DATE ________

Available as a transparency

# Calculator Lesson Opener

For use with pages 465–471

**1.** The *geometric mean* of two positive numbers $a$ and $b$ is the positive number $x$ such that $\frac{a}{x} = \frac{x}{b}$. Solve this proportion for $x$.

**2.** Use a calculator to complete the table for the proportion $\frac{a}{x} = \frac{x}{b}$.

| $a$ | $b$ | $x$ |
|---|---|---|
| 5 | 20 | |
| 2 | 13 | |
| 6 | 8 | |
| 12 | 3 | |
| 11 | 11 | |
| 1 | 9 | |
| 30 | 7 | |
| 9 | 12 | |
| | 8 | 4 |
| 2 | | 8 |
| | 25 | 5 |
| 14 | | 14 |
| | 5 | 15 |
| 4 | | 6 |

**3.** What must be true of the product $a \cdot b$ if $x$ is a whole number?

**4.** Find eight different pairs $a$ and $b$ with a geometric mean of 12.

**5.** Find a whole-number value of $x$ such that there are exactly two different pairs $a$ and $b$ with a geometric mean of $x$. What must be true of $x$?

**6.** Find a whole-number value of $x$ such that there are exactly five different pairs $a$ and $b$ with a geometric mean of $x$. Write the pairs.

NAME _____ DATE _____

## Practice A

**For use with pages 465–471**

**Complete the sentence.**

1. If $\dfrac{a}{b} = \dfrac{3}{4}$, then $\dfrac{b}{a} = \dfrac{?}{?}$.

2. If $\dfrac{a}{b} = \dfrac{3}{4}$, then $\dfrac{a}{3} = \dfrac{?}{?}$.

3. If $\dfrac{a}{b} = \dfrac{3}{4}$, then $\dfrac{a+b}{b} = \dfrac{?}{?}$.

4. If $\dfrac{a}{b} = \dfrac{3}{4}$, then $\dfrac{?}{?} = \dfrac{7}{4}$.

**Decide whether the statement is *true* or *false*.**

5. If $\dfrac{m}{n} = \dfrac{4}{5}$, then $\dfrac{n}{m} = \dfrac{4}{5}$.

6. If $\dfrac{m}{n} = \dfrac{3}{6}$, then $\dfrac{3}{n} = \dfrac{m}{6}$.

7. If $\dfrac{m}{n} = \dfrac{2}{3}$, then $\dfrac{m+n}{n} = \dfrac{5}{3}$.

8. If $\dfrac{m}{n} = \dfrac{3}{4}$, then $\dfrac{m-n}{n} = -\dfrac{1}{4}$.

**Find the geometric mean of the two numbers.**

9. 4 and 9

10. 4 and 16

11. 3 and 12

12. 5 and 20

13. 4 and 8

14. 6 and 12

**Use the diagram and the given information to find the unknown length.**

15. Given: $\dfrac{AB}{BD} = \dfrac{AC}{CE}$, find $BD$.

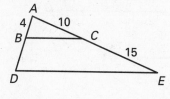

16. Given: $\dfrac{MN}{NO} = \dfrac{MP}{PQ}$, find $PQ$.

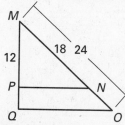

**In Exercises 17 and 18, construct a verbal model and solve the proportion.**

17. The recommended application for a particular type of lawn fertilizer is one 50-pound bag for 575 square feet. How many bags of this type of fertilizer would be required to fertilize 2850 square feet of lawn?

   **Verbal Model:** $\dfrac{\text{a. }?}{\text{b. }?} = \dfrac{\text{c. }?}{\text{d. }?}$

18. You have just moved into a new neighborhood and a new house valued at $110,000. If your next door neighbor pays $1,150 in real estate taxes each year on a house valued at $89,000, how much a year should you expect to pay in real estate taxes? (Assume that the rate is the same.)

   **Verbal Model:** $\dfrac{\text{a. }?}{\text{b. }?} = \dfrac{\text{c. }?}{\text{d. }?}$

Lesson 8.2

## Practice B

For use with pages 465–471

**Complete the sentence.**

**1.** If $\dfrac{p}{q} = \dfrac{5}{8}$, then $\dfrac{q}{p} = \dfrac{?}{?}$.

**2.** If $\dfrac{p}{q} = \dfrac{5}{8}$, then $\dfrac{p}{5} = \dfrac{?}{?}$.

**3.** If $\dfrac{p}{q} = \dfrac{5}{8}$, then $\dfrac{p+q}{q} = \dfrac{?}{?}$.

**4.** If $\dfrac{p}{q} = \dfrac{5}{8}$, then $\dfrac{?}{?} = \dfrac{13}{8}$.

**Decide whether the statement is *true* or *false*.**

**5.** If $\dfrac{x}{y} = \dfrac{2}{9}$, then $\dfrac{y}{x} = \dfrac{9}{2}$.

**6.** If $\dfrac{x}{y} = \dfrac{2}{9}$, then $\dfrac{2}{y} = \dfrac{x}{9}$.

**7.** If $\dfrac{x}{y} = \dfrac{2}{9}$, then $\dfrac{9}{y} = \dfrac{2}{x}$.

**8.** If $\dfrac{x}{y} = \dfrac{2}{9}$, then $\dfrac{x-y}{y} = \dfrac{7}{9}$.

**Find the geometric mean of the two numbers.**

**9.** 6 and 10

**10.** 8 and 12

**11.** 5 and 24

**12.** 10 and 15

**13.** 12 and 16

**14.** 20 and 24

**Use the diagram and the given information to find the unknown length.**

**15.** Given: $\dfrac{LJ}{JN} = \dfrac{MK}{KP}$, find $JN$.

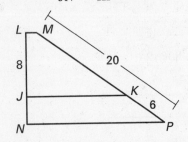

**16.** Given: $\dfrac{MN}{NO} = \dfrac{MP}{PQ}$, find $PQ$.

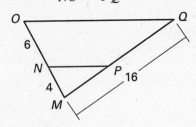

**17.** In December 1999, the exchange rate of Mexican pesos to American dollars was 9.52 to 1. You paid 450 pesos for a jacket. Use the following verbal model to find the price of the jacket in dollars.

$$\dfrac{\text{Price in pesos}}{\text{Price in dollars}} = \dfrac{9.52 \text{ pesos}}{1 \text{ dollar}}$$

**18.** In December 1999, the exchange rate of Canadian dollars to American dollars was 1 to 0.68. You paid $30.00 (in Canadian dollars) for a sweater. What was the price of the sweater in American dollars?

**19.** The Wright brothers made the world's first flight in a power-driven airplane. The flight lasted for 12 seconds at an average speed of 10 feet per second. The ratio of the airplane's wingspan to the distance flown was 1:3. How long was the wingspan?

Lesson 8.2

NAME _____ DATE _____

## Practice C

**For use with pages 465–471**

**Complete the sentence.**

1. If $\dfrac{m}{n} = \dfrac{5}{9}$, then $\dfrac{n}{m} = \dfrac{?}{?}$.

2. If $\dfrac{m}{n} = \dfrac{5}{9}$, then $\dfrac{m}{5} = \dfrac{?}{?}$.

3. If $\dfrac{m}{n} = \dfrac{5}{9}$, then $\dfrac{m+n}{n} = \dfrac{?}{?}$.

4. If $\dfrac{m}{n} = \dfrac{5}{9}$, then $\dfrac{?}{?} = \dfrac{14}{9}$.

**Find the geometric mean of the two numbers.**

5. 8 and 12

6. 8.5 and 12.4

7. 15 and 24

8. 18 and 30

9. $a$ and $4a$

10. $2a$ and $4a$

**Use the diagram and the given information to find the unknown length.**

11. **Given:** $\dfrac{AB}{AC} = \dfrac{DE}{DF}$, find $EF$.

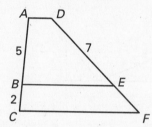

12. **Given:** $\dfrac{JK}{KL} = \dfrac{JM}{MN}$, find $JN$.

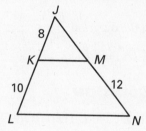

13. The points $(-2, -3)$, $(8, 7)$, and $(x, -6)$ are collinear. Find the value of $x$ by solving the proportion below.

$$\dfrac{(-3) - 7}{(-2) - 8} = \dfrac{(-3) - (-6)}{-2 - x}$$

14. The points $(-4, 6)$, $(2, -2)$, and $(x, -6)$ are collinear. Find the value of $x$ by solving the proportion below.

$$\dfrac{6 - (-2)}{(-4) - 2} = \dfrac{-2 - (-6)}{2 - x}$$

15. A quality control engineer for a certain buyer found that the ratio of defective units to total units is 1:35. At this rate, what is the expected number of defective units in a shipment of 28,000?

16. The scale represents 100 miles on the accompanying map. Approximate the distance between Philadelphia and Pittsburgh.

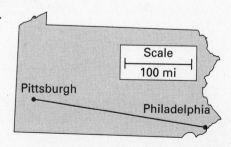

NAME _____ DATE _____

# Reteaching with Practice

For use with pages 465–471

**GOAL** **Use properties of proportions**

---

### VOCABULARY

The **geometric mean** of two positive numbers $a$ and $b$ is the positive number $x$ such that $\dfrac{a}{x} = \dfrac{x}{b}$.

**Additional Properties of Proportions**

3. If $\dfrac{a}{b} = \dfrac{c}{d}$, then $\dfrac{a}{c} = \dfrac{b}{d}$

4. If $\dfrac{a}{b} = \dfrac{c}{d}$, then $\dfrac{a+b}{b} = \dfrac{c+d}{d}$

---

**EXAMPLE 1** *Using Properties of Proportions*

In the diagram $\dfrac{BC}{DC} = \dfrac{AC}{EC}$. Find the lengths of $\overline{DC}$ and $\overline{BC}$.

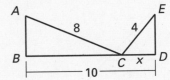

**SOLUTION**

Let $DC = x$. Then $BC = 10 - x$.

$\dfrac{BC}{DC} = \dfrac{AC}{EC}$      Given

$\dfrac{10 - x}{x} = \dfrac{8}{4}$      Substitute.

$\dfrac{10 - x}{x} = \dfrac{2}{1}$      Simplify.

$10 - x = 2x$      Cross product property

$10 = 3x$      Add $x$ to each side.

$\dfrac{10}{3} = x$      Divide each side by 3.

$x \approx 3.3$      Use a calculator.

So, $DC \approx 3.3$ and $BD \approx 10 - 3.3 = 6.7$.

---

NAME _____ DATE _____

# Reteaching with Practice

For use with pages 465–471

### Exercises for Example 1

**Find the value of *x*.**

1. $\dfrac{AB}{BC} = \dfrac{FE}{ED}$

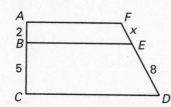

2. $\dfrac{BG}{CD} = \dfrac{AB}{AC}$

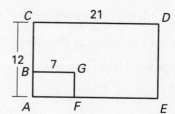

**EXAMPLE 2** *Using a Geometric Mean*

In the diagram $\dfrac{AC}{DC} = \dfrac{DC}{BC}$. Find the value of *x*.

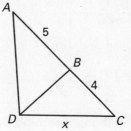

#### SOLUTION

Let $DC = x$.

$\dfrac{AC}{x} = \dfrac{x}{BC}$      Write proportion.

$\dfrac{9}{x} = \dfrac{x}{4}$      Substitute.

$x^2 = 9 \cdot 4$      Cross product property

$x = \sqrt{36} = 6$      Simplify.

### Exercises for Example 2

**Find the value of *x*.**

3. $\dfrac{CD}{AB} = \dfrac{AB}{EF}$

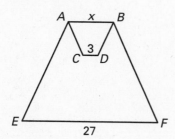

4. $\dfrac{AB}{AD} = \dfrac{AD}{DC}$

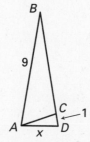

NAME _____ DATE _____

# Quick Catch-Up for Absent Students

**For use with pages 465–471**

The items checked below were covered in class on (date missed) _____

**Lesson 8.2: Problem Solving in Geometry with Proportions**

____ **Goal 1:** Use properties of proportions. (pp. 465–466)

*Material Covered:*

____ Example 1: Using Properties of Proportions

____ Example 2: Using Properties of Proportions

____ Student Help: Skills Review

____ Example 3: Using a Geometric Mean

*Vocabulary:*

geometric mean, p. 466

____ **Goal 2:** Use proportions to solve real-life problems. (p. 467)

*Material Covered:*

____ Example 4: Solving a Proportion

____ Other (specify) _____

_____

**Homework and Additional Learning Support**

____ Textbook (specify) _pp. 468–471_____

_____

____ Internet: Extra Examples at www.mcdougallittell.com

____ *Reteaching with Practice* worksheet (specify exercises)_____

____ *Personal Student Tutor* for Lesson 8.2

# *Interdisciplinary Application*

For use with pages 465–471

## Map Making

**GEOGRAPHY**   Cartography is the study of map making. We have knowledge of map making dating back to the early Babylonians. These maps were little more than rough descriptions and relative positions of landmarks familiar to the cartographer. Map making did not start to take a modern form until the Greeks. Aristotle was one of the ancient scholars who argued through observation and logic that Earth is a sphere, and Dicaearchus, a disciple of Aristotle, first used lines of reference which were precursors to longitude and latitude lines. The Greeks drew on information gathered by Phoenecian sailors and others to make more accurate maps based on relative position and proportional distances. During the age of Discovery and Exploration in the 15th century, Europeans used their new technology in ship building, compass making and other navigation innovations to create more comprehensive and accurate global maps.

Modern maps are accurate reproductions of the true shape and size of the world. Advanced surveying techniques and aerial observations allow us to make more precise maps. These maps are proportional reproductions of this new data we are able to gather.

### In Exercises 1–5, use the area of each of four Mid-Atlantic states given below.

North Carolina: 53,821 $mi^2$, or 139,391 $km^2$

South Carolina: 31,055 $mi^2$, or 80,432 $km^2$

Maryland: 9837 $mi^2$, or 25,476.85 $km^2$

Virginia: 40,817 $mi^2$, or 105,716 $km^2$

1. If you were going to cut an image of Virginia out of a piece of cardboard, how many square inches would you need if the scale was 1 square inch for every 100 square miles (1:100)?

2. How large would the cardboard have to be in order to cut out all of the Mid-Atlantic states?

3. If 20% of a square piece of cardboard were wasted because the shapes of the states do not fit precisely, how large (in square inches) would the cardboard square have to be in order to cut out all of the Mid-Atlantic states? (*Hint:* Total area of states = 80% of total area of cardboard square.)

4. What are the dimensions of the cardboard square in Exercise 3? Round your answer to the nearest tenth of an inch.

5. Set up a proportion and find the scale needed to fit the shapes onto a square piece of cardboard that is 15 inches by 15 inches. Use the same 20% figure for waste and round to the nearest tenth.

NAME _____ DATE _____

# Challenge: Skills and Applications

For use with pages 465–471

**In Exercises 1–6, use the given information to find all possible values of $x$. (Assume the given quantities must be positive.)**

1. The geometric mean of $x - 3$ and $x + 4$ is $x$.

2. The geometric mean of $x$ and $x^2$ is 8.

3. The geometric mean of $x + 1$ and $12x$ is $6x$.

4. The geometric mean of $\sqrt{x}$ and $9\sqrt{x}$ is $x - 4$.

5. The geometric mean of $x - 3$ and $2x + 8$ is $x + 4$.

6. The geometric mean of $x + 1$ and $3x + 1$ is $3x - 1$.

**In Exercises 7–9, give each answer in terms of $x$.**

7. Given: $\dfrac{AB}{BC} = \dfrac{FE}{ED}$, find $FE$.

8. Given: $\dfrac{GH}{GJ} = \dfrac{GI}{GK}$, find $HJ$.

9. Given: $\dfrac{MN}{LN} = \dfrac{PN}{QN}$, find $LN$.

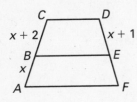

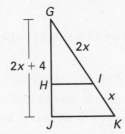

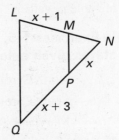

**In Exercises 10–12, use the given information to find all possible values of $x$.**

10. Given: $\dfrac{EG}{GI} = \dfrac{FH}{HI}$

11. Given: $\dfrac{ST}{RT} = \dfrac{UV}{RV}$

12. Given: $\dfrac{JK}{KL} = \dfrac{JM}{MN}$

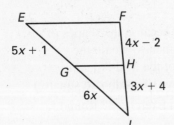

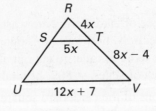

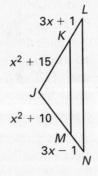

13. An airplane has a wingspan of $(x^2 + 1)$ ft and a length of $(x^2 - 9)$ ft. A scale model of this plane has a wingspan of $(x + 3)$ ft and a length of $(x + 1)$ ft. Based on this information, use a proportion to find the wingspan of the actual airplane.

TEACHER'S NAME _____ CLASS _____ ROOM _____ DATE _____

# *Lesson Plan*

2-day lesson (See *Pacing the Chapter,* TE pages 454C–454D)          **For use with pages 472–479**

**GOALS**   1. **Identify similar polygons.**
2. **Use similar polygons to solve real-life problems.**

State/Local Objectives _____

_____

## ✓ Check the items you wish to use for this lesson.

### STARTING OPTIONS
____ Homework Check: TE page 468: Answer Transparencies
____ Warm-Up or Daily Homework Quiz: TE pages 473 and 471, CRB page 35, or Transparencies

### TEACHING OPTIONS
____ Concept Activity: SE page 472; CRB page 36 (Activity Support Master)
____ Lesson Opener (Visual Approach): CRB page 37 or Transparencies
____ Examples:   Day 1: 1–4, SE pages 473–474; Day 2: 5, SE page 475
____ Extra Examples:   Day 1: TE page 474 or Transp.; Day 2: TE page 475 or Transp.
____ Closure Question: TE page 475
____ Guided Practice: SE page 475   Day 1: Exs. 1–7; Day 2: none

### APPLY/HOMEWORK
#### Homework Assignment
____ Basic   Day 1: 8–38; Day 2: 39–42, 45–49, 53–67; Quiz 1: 1–10
____ Average   Day 1: 8–38; Day 2: 39–42, 45–49, 53–67; Quiz 1: 1–10
____ Advanced   Day 1: 8–38; Day 2: 39–42, 45–49, 50–67; Quiz 1: 1–10

#### Reteaching the Lesson
____ Practice Masters: CRB pages 38–40 (Level A, Level B, Level C)
____ Reteaching with Practice: CRB pages 41–42 or Practice Workbook with Examples
____ Personal Student Tutor

#### Extending the Lesson
____ Applications (Real-Life): CRB page 44
____ Challenge: SE page 478; CRB page 45 or Internet

### ASSESSMENT OPTIONS
____ Checkpoint Exercises:   Day 1: TE page 475 or Transp.; Day 2: TE page 475 or Transp.
____ Daily Homework Quiz (8.3): TE page 479, CRB page 49, or Transparencies
____ Standardized Test Practice: SE page 478; TE page 479; STP Workbook; Transparencies
____ Quiz (8.1–8.3): SE page 479; CRB page 46

Notes _____

_____

_____

TEACHER'S NAME _____ CLASS _____ ROOM _____ DATE _____

# Lesson Plan for Block Scheduling

1-day lesson (See *Pacing the Chapter*, TE pages 454C–454D)          For use with pages 472–479

**GOALS**   1. **Identify similar polygons.**
2. **Use similar polygons to solve real-life problems.**

State/Local Objectives _____

_____

_____

✓ **Check the items you wish to use for this lesson.**

### STARTING OPTIONS
_____ Homework Check: TE page 468: Answer Transparencies
_____ Warm-Up or Daily Homework Quiz: TE pages 473 and
      471, CRB page 35, or Transparencies

### TEACHING OPTIONS
_____ Concept Activity: SE page 472; CRB page 36 (Activity Support Master)
_____ Lesson Opener (Visual Approach): CRB page 37 or Transparencies
_____ Examples:  Day 2: 1–4, SE pages 473–474; Day 3: 5, SE page 475
_____ Extra Examples:  Day 2: TE page 474 or Transp.; Day 3: TE page 475 or Transp.
_____ Closure Question: TE page 475
_____ Guided Practice: SE page 475   Day 2: Exs. 1–7; Day 3: none

### APPLY/HOMEWORK
**Homework Assignment  (See also the assignments for Lessons 8.2 and 8.4.)**
_____ Block Schedule:  Day 2: 8–38; Day 3: 39–42, 45–49, 53–67; Quiz 1: 1–10

### Reteaching the Lesson
_____ Practice Masters: CRB pages 38–40 (Level A, Level B, Level C)
_____ Reteaching with Practice: CRB pages 41–42 or Practice Workbook with Examples
_____ Personal Student Tutor

### Extending the Lesson
_____ Applications (Real-Life): CRB page 44
_____ Challenge: SE page 478; CRB page 45 or Internet

### ASSESSMENT OPTIONS
_____ Checkpoint Exercises:  Day 2: TE page 475 or Transp.; Day 3: TE page 475 or Transp.
_____ Daily Homework Quiz (8.3): TE page 479, CRB page 49, or Transparencies
_____ Standardized Test Practice: SE page 478; TE page 479; STP Workbook; Transparencies
_____ Quiz (8.1–8.3): SE page 479; CRB page 46

| CHAPTER PACING GUIDE | |
| --- | --- |
| Day | Lesson |
| 1 | Assess Ch. 7: 8.1 (all) |
| 2 | 8.2 (all); **8.3 (begin)** |
| 3 | **8.3 (end)**; 8.4 (begin) |
| 4 | 8.4 (end); 8.5 (begin) |
| 5 | 8.5 (end); 8.6 (begin) |
| 6 | 8.6 (end); 8.7 (begin) |
| 7 | 8.7 (end); Review Ch. 8 |
| 8 | Assess Ch. 8; 9.1 (all) |

Notes _____

_____

_____

_____

NAME _____ DATE _____

# WARM-UP EXERCISES

For use before Lesson 8.3, pages 472–479

**Find the measure of the angles in the figure below.**

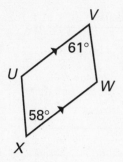

**1.** $\angle U$

**2.** $\angle W$

**Solve each proportion.**

**3.** $\dfrac{2}{x} = \dfrac{5}{20}$

**4.** $\dfrac{18}{5} = \dfrac{12}{t}$

# DAILY HOMEWORK QUIZ

For use after Lesson 8.2, pages 465–471

**Decide whether the statement is *true* or *false*.**

**1.** If $\dfrac{a}{6} = \dfrac{b-1}{8}$, then $\dfrac{a+6}{6} = \dfrac{b+8}{8}$.

**2.** If $\dfrac{x}{y} = \dfrac{x+5}{y-3}$, then $\dfrac{x}{x+5} = \dfrac{y}{y-3}$.

**3.** Find the geometric mean of 9 and 25.

**4.** Given: $\dfrac{PT}{PR} = \dfrac{QU}{QS}$, find $SU$.

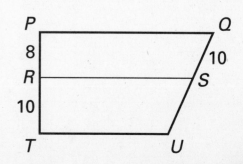

LESSON 8.3

NAME ______________________________ DATE __________

# Activity Support Master

For use with page 472

| Measurement | Photo 1 | Photo 2 | Ratio |
|---|---|---|---|
| *AB* | 4.2 cm | 3.0 cm | $\frac{4.2}{3.0} = 1.4$ |
| *AF* | | | |
| *CD* | | | |
| $m\angle 1$ | | | |
| $m\angle 2$ | | | |
| Perimeter of photo | | | |
| | | | |
| | | | |
| | | | |
| | | | |
| | | | |

Lesson 8.3

# *Visual Approach Lesson Opener*

For use with pages 473–479

The Lute of Pythagoras is shown twice below.

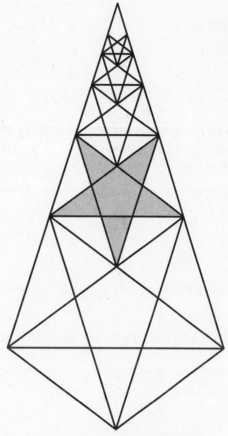

1. Color the Lute of Pythagoras on the left so that all figures that are congruent are the same color. Then find the five pentagons that are *similar* (have the same shape) and number them from smallest to largest.

2. Color the Lute of Pythagoras on the right so that each star similar to the shaded star is a different color.

3. Four different shapes make up the Lute of Pythagoras: pentagon, obtuse isosceles triangle, acute isosceles triangle, and kite. How many of each shape are there? (*Hint:* The total should be 51.)

*Lesson 8.3*

NAME _____ DATE _____

## *Practice A*

For use with pages 473–479

**You are given the length and width of three rectangles. Which two are similar?**

**1. a.** 5 in. × 7 in.          **b.** 8.5 in. × 11 in.          **c.** 10 in. × 14 in.

**2. a.** 4 ft × 5 ft          **b.** 20 ft × 25 ft          **c.** 8 cm × 1 m

**3. a.** 3 cm × 15 cm          **b.** 3 ft × 15 in.          **c.** 6 cm × 30 cm

**4. a.** $\frac{3}{2}$ cm × $\frac{7}{2}$ cm          **b.** 21 in. × 49 in.          **c.** 1 ft × 3 ft

**List all pairs of congruent angles and write the statement of proportionality for the figures.**

**5.** $\triangle GRM \sim \triangle TFD$

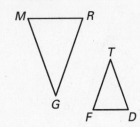

**6.** $\triangle STR \sim \triangle JKL$

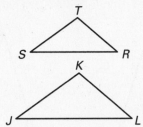

**7.** $\square ABCD \sim \square MNOP$

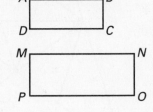

**Decide whether the polygons are similar. If so, write a similarity statement.**

**8.**

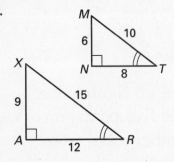

**9.**

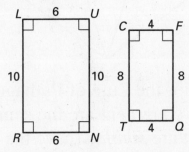

**In the diagram at the right, polygon *ABCD* ~ polygon *GHIJ*.**

**10.** Find the scale factor of polygon *ABCD* to polygon *GHIJ*.

**11.** Find the scale factor of polygon *GHIJ* to polygon *ABCD*.

**12.** Find the values of *x* and *y*.

**13.** Find the perimeter of each polygon.

**14.** Find the ratio of the perimeter of *ABCD* to the perimeter of *GHIJ*.

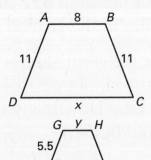

NAME _____ DATE _____

# Practice B

For use with pages 473–479

**List all pairs of congruent angles and write the statement of proportionality for the figures.**

**1.** △JKL ~ △RST

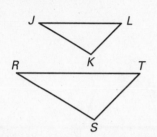

**2.** △TUV ~ △MNO

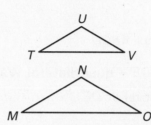

**3.** ▱WXYZ ~ ▱DEFG

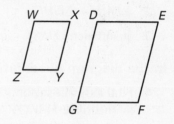

**Decide whether the polygons are similar. If so, write a similarity statement.**

**4.**

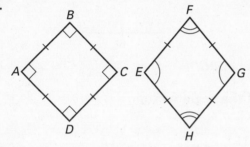

**5.**

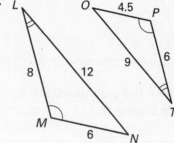

**In the diagram at the right, ▱RSTU ~ ▱LMNO.**

**6.** Find the scale factor of ▱RSTU to ▱LMNO.

**7.** Find the scale factor of ▱LMNO to ▱RSTU.

**8.** Find the length of $\overline{NO}$.

**9.** Find the measure of ∠U.

**10.** Find the perimeter of ▱LMNO.

**11.** Find the ratio of the perimeter of ▱RSTU to the perimeter of ▱LMNO.

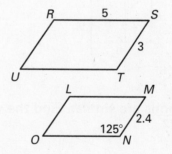

**The two polygons are similar. Find the values of x and y.**

**12.**

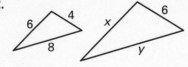

**13.**

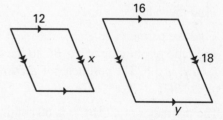

**14.** The ratio of one side of △ABC to the corresponding side of similar △DEF is 3:5. The perimeter of △DEF is 48 inches. What is the perimeter of △ABC?

NAME _____ DATE _____

## *Practice C*
For use with pages 473–479

**List all pairs of congruent angles and write the statement of pro-portionality for the figures.**

1. $\triangle STU \sim \triangle CDE$

2. $\triangle LMN \sim \triangle GHI$

3. quadrilateral $QRST \sim$ quadrilateral $ABCD$

**In the diagram quadrilateral *BCDE* ~ quadrilateral *WXYZ*.**

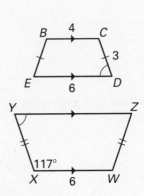

4. Find the scale factor of quadrilateral $BCDE$ to quadrilateral $WXYZ$.

5. Find the scale factor of quadrilateral $WXYZ$ to quadrilateral $BCDE$.

6. Find the length of $\overline{XY}$.

7. Find the measure of $\angle D$.

8. Find the perimeter of quadrilateral $WXYZ$.

9. Find the ratio of the perimeter of $WXYZ$ to the perimeter of $BCDE$.

**Decide whether the polygons are similar. If so, find the scale factor of Figure A to Figure B.**

10.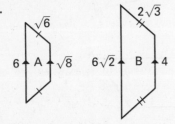

11.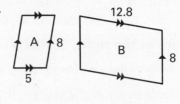

**The two polygons are similar. Find the values of *x* and *y*.**

12.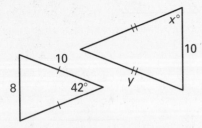

13.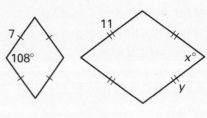

14. The ratio of one side of $\triangle ABC$ to the corresponding side of similar $\triangle DEF$ is 5:8. The perimeter of $\triangle DEF$ is 96 inches. What is the perimeter of $\triangle ABC$?

15. The perimeter of $\square ABCD$ is 60 centimeters. The perimeter of $\square EFGH$ is 15 centimeters and $\square ABCD \sim \square EFGH$. The lengths of two of the sides of $\square ABCD$ are 18 centimeters each. Find the scale factor of $\square ABCD$ to $\square EFGH$, and the lengths of the sides of $\square EFGH$.

NAME _____ DATE _____

# Reteaching with Practice

For use with pages 473–479

**GOAL** Identify and use similar polygons

### VOCABULARY

When there is a correspondence between two polygons such that their corresponding angles are congruent and the lengths of corresponding sides are proportional the two polygons are called **similar polygons.**

**Theorem 8.1**   If two polygons are similar, then the ratio of their perimeters is equal to the ratios of their corresponding side lengths.

**EXAMPLE 1** *Writing Similarity Statements*

Quadrilaterals *ABCD* and *EFGH* are similar. List all the pairs of congruent angles. Write the ratios of the corresponding sides in a statement of proportionality.

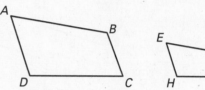

### SOLUTION

Because *ABCD* ~ *EFGH* you can write$\angle A \cong \angle E$, $\angle B \cong \angle F$, $\angle C \cong \angle G$, and $\angle D \cong \angle H$. You can write the statement of proportionality as follows:

$$\frac{AB}{EF} = \frac{BC}{FG} = \frac{CD}{GH} = \frac{DA}{HE}.$$

### Exercises for Example 1

**The two polygons are similar. List all the pairs of congruent angles. Write the ratios of the corresponding sides in a statement of proportionality.**

**1.** $\triangle ABC \sim \triangle DEF$

**2.** $ABDC \sim ZWXY$

**3.** $EFGHJ \sim MRQPN$

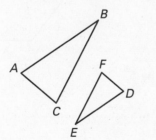

**EXAMPLE 2** *Comparing Similar Polygons*

Decide whether the figures are similar. If they are similar, write a similarity statement.

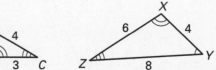

Lesson 8.3

# Reteaching with Practice

For use with pages 473–479

## SOLUTION

The corresponding angles of $\triangle ABC$ and $\triangle XYZ$ are congruent. Also, the corresponding side lengths are proportional.

$$\frac{AB}{XY} = \frac{2}{4} = \frac{1}{2} \qquad\qquad \frac{BC}{YZ} = \frac{4}{8} = \frac{1}{2} \qquad\qquad \frac{CA}{ZX} = \frac{3}{6} = \frac{1}{2}$$

So, the two triangles are similar and you can write $\triangle ABC \sim \triangle XYZ$.

## Exercises for Example 2

**Are the polygons similar? If so, write a similarity statement.**

**4.**

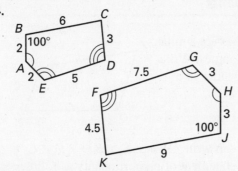

**5.**

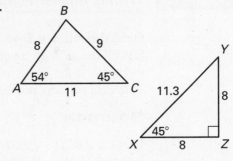

---

**EXAMPLE 3** | *Using Similar Polygons*

Pentagon *ABCDE* is similar to
pentagon *JKLMN*. Find the value of *x*.

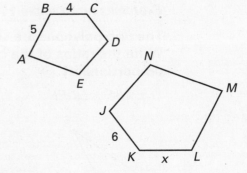

### SOLUTION

Set up a proportion that contains *KL*.

$\dfrac{AB}{JK} = \dfrac{BC}{KL}$ \qquad Write a proportion.

$\dfrac{5}{6} = \dfrac{4}{x}$ \qquad Substitute.

$x = 4.8$ \qquad Cross multiply and divide by 5.

## Exercises for Example 3

**Find the value of x.**

**6.** *ABCD ~ WXYZ*

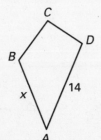

**7.** *JKLMN ~ PQRST*

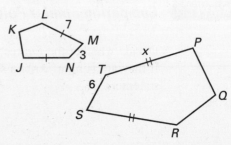

**Geometry**
Chapter 8 Resource Book

NAME _____ DATE _____

# Quick Catch-Up for Absent Students

For use with pages 472–479

The items checked below were covered in class on (date missed) _____

**Activity 8.3: Making Conjectures about Similarity (p. 472)**

_____ **Goal:** Make a conjecture about how corresponding areas are related when a figure is enlarged.

**Lesson 8.3: Similar Polygons**

_____ **Goal 1:** Identify similar polygons. (p. 473)

*Material Covered:*

_____ Example 1: Writing Similarity Statements

_____ Student Help: Study Tip

_____ Example 2: Comparing Similar Polygons

*Vocabulary:*

similar polygons, p. 473

_____ **Goal 2:** Use similar polygons to solve real-life problems. (pp. 474–475)

*Material Covered:*

_____ Example 3: Comparing Photographic Enlargements

_____ Example 4: Using Similar Polygons

_____ Example 5: Using Similar Polygons

*Vocabulary:*

scale factor, p. 474

_____ Other (specify) _____

_____

**Homework and Additional Learning Support**

_____ Textbook (specify) _pp. 475–479_____

_____

_____ *Reteaching with Practice* worksheet (specify exercises)_____

_____ *Personal Student Tutor* for Lesson 8.3

NAME _____ DATE _____

# Real-Life Application: When Will I Ever Use This?

For use with pages 473–479

## Architecture

An architect uses many methods to help visualize the building structure. This includes the placement of doors, windows, and even landscaping. A plan starts at the drafting board with a two-dimensional drawing. From there, an architect can move into three-dimensional models. These models are true to scale, including landscaping and surrounding buildings. These models are used for sales, presentations and displays.

In today's technology, an architect uses CAD, virtual reality, and animation to display a building design. Using computer software, an architect can produce three-dimensional models and then manipulate this model in a variety of ways. This also allows the architect to change the dimensions or the design of the building. Virtual reality software can display a "walk through" model of the building before construction even begins.

### In Exercises 1–4, use the following information.

An architect is designing an office building that is to be a width of 50 feet, a length of 50 feet, and a height of 100 feet. A scale model of the same building is to have a width of 18 inches, a length of 18 inches, and a height of 36 inches.

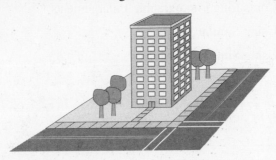

1. Find the scale factor of the office building to the model.

2. The entrance into the office building is 8 feet high by 7 feet wide. Find the height and the width of the entrance on the scale model. Round your answer to one decimal place.

3. Two windows on either side of the main entrance measure 3 inches wide and 2 inches high on the scale model. Find the height and the width on the actual building. Round your answer to one decimal place.

4. You want to add a window over the entrance to the building. Determine the size that you want the actual window to be. Then find the size of the window on the scale model.

**Geometry**
Chapter 8 Resource Book

# Challenge: Skills and Applications

**For use with pages 473–479**

1. In the diagram, *PQSR* is a square and
   *RSUT* ~ *TPQU*. Find the value of *x*.
   Express your answer in exact form and
   as a decimal approximation. (This
   number is known as the *golden ratio*.)

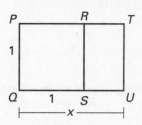

2. In the diagram, *KLMN* ~ *WXYZ*.

   **a.** Find *WX*, *XY*, and *YZ* in terms of *r*, *s*, *t*, *u*,
   and *v*.

   **b.** Use the result of part (a) to show that the
   ratio of the perimeters is the same as the
   ratio of any pair of corresponding sides.

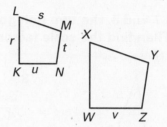

3. In the diagram, $\overline{BA} \parallel \overline{CD}$ and $\dfrac{CD}{AB} = \dfrac{DE}{BC}$. Write
   a paragraph proof to show that $\triangle ABC \sim \triangle CDE$.
   (Hint: Let $k = \dfrac{CD}{AB}$. You may use the Pythagorean
   Theorem.)

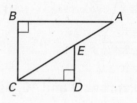

**In Exercises 4 and 5, the two triangles are similar.
Find all possible values of *x*.**

4. **Given:** $\triangle FGH \sim \triangle JKL$

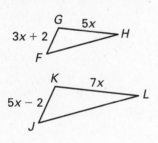

5. **Given:** $\triangle PQR \sim \triangle STU$

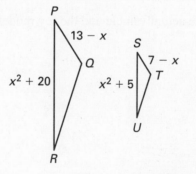

6. A 5-inch by 8-inch photo was enlarged to make
   a poster, as shown. If the dimensions of the
   poster are $(x^2 - 6)$ inches by $(x^2 + 12)$ inches,
   what is the area of the poster?

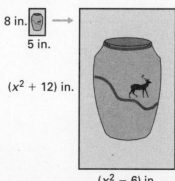

NAME _____ DATE _____

## Quiz 1

For use after Lessons 8.1–8.3

**Solve the proportions.** *(Lesson 8.1)*

**1.** $\dfrac{x}{4} = \dfrac{9}{2}$        **2.** $\dfrac{3}{4} = \dfrac{6}{p}$        **3.** $\dfrac{5}{3x-2} = \dfrac{10}{x}$

**Find the geometric mean of the two numbers.** *(Lesson 8.2)*

**4.** 5 and 125       **5.** 8 and 9       **6.** 12 and 9

**In Exercises 7 and 8, the two polygons are similar. Find the value of x. Then find the scale factor and the ratio of the perimeters.** *(Lesson 8.3)*

**7.**

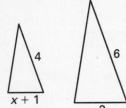

**8.**

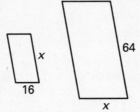

**Answers**

1. _____

2. _____

3. _____

4. _____

5. _____

6. _____

7. _____

8. _____

9. _____

10. _____

*Toy Model* **In Exercises 9 and 10, use the following information.** *(Lesson 8.3)*

You bought a toy model of a sports utility vehicle for your little brother. The scale factor of the actual sports utility vehicle to the toy model is 60 : 1.

**9.** Are the toy model and actual vehicle similar?

**10.** Do the angles on the actual vehicle and the toy model have a ratio of 60 : 1?

TEACHER'S NAME _____ CLASS _____ ROOM _____ DATE _____

# *Lesson Plan*

2-day lesson (See *Pacing the Chapter*, TE pages 454C–454D)          For use with pages 480–487

**GOALS**   1. **Identify similar triangles.**
2. **Use similar triangles in real-life problems.**

State/Local Objectives _____

_____

## ✓ Check the items you wish to use for this lesson.

**STARTING OPTIONS**
____ Homework Check: TE page 476: Answer Transparencies
____ Warm-Up or Daily Homework Quiz: TE pages 480 and 479, CRB page 49, or Transparencies

**TEACHING OPTIONS**
____ Lesson Opener (Application): CRB page 50 or Transparencies
____ Technology Activity with Keystrokes: CRB pages 51–54
____ Examples:   Day 1: 1–3, SE pages 480–481; Day 2: 4–5, SE page 482
____ Extra Examples:   Day 1: TE page 481 or Transp.; Day 2: TE page 482 or Transp.; Internet
____ Closure Question: TE page 482
____ Guided Practice: SE page 483   Day 1: Exs. 1–8; Day 2: none

**APPLY/HOMEWORK**
**Homework Assignment**
____ Basic   Day 1: 9–38; Day 2: 39–52, 57, 61–71
____ Average   Day 1: 9–38; Day 2: 39–53, 55–57, 61–71
____ Advanced   Day 1: 9–38; Day 2: 39–53, 55–71

**Reteaching the Lesson**
____ Practice Masters: CRB pages 55–57 (Level A, Level B, Level C)
____ Reteaching with Practice: CRB pages 58–59 or Practice Workbook with Examples
____ Personal Student Tutor

**Extending the Lesson**
____ Applications (Interdisciplinary): CRB page 61
____ Challenge: SE page 487; CRB page 62 or Internet

**ASSESSMENT OPTIONS**
____ Checkpoint Exercises:   Day 1: TE page 481 or Transp.; Day 2: TE page 482 or Transp.
____ Daily Homework Quiz (8.4): TE page 487, CRB page 65, or Transparencies
____ Standardized Test Practice: SE page 487; TE page 487; STP Workbook; Transparencies

Notes _____

_____

_____

TEACHER'S NAME _____ CLASS _____ ROOM _____ DATE _____

# Lesson Plan for Block Scheduling

1-day lesson (See *Pacing the Chapter,* TE pages 454C–454D)          For use with pages 480–487

**GOALS**   1. **Identify similar triangles.**
           2. **Use similar triangles in real-life problems.**

State/Local Objectives _____

_____

_____

✓ **Check the items you wish to use for this lesson.**

## STARTING OPTIONS

____ Homework Check: TE page 476: Answer Transparencies
____ Warm-Up or Daily Homework Quiz: TE pages 480 and
         479, CRB page 49, or Transparencies

## TEACHING OPTIONS

____ Lesson Opener (Application): CRB page 50 or Transparencies
____ Technology Activity with Keystrokes: CRB pages 51–54
____ Examples:   Day 3: 1–3, SE pages 480–481; Day 4: 4–5, SE page 482
____ Extra Examples:   Day 3: TE page 481 or Transp.; Day 4: TE page 482 or Transp.; Internet
____ Closure Question: TE page 482
____ Guided Practice: SE page 483   Day 3: Exs. 1–8; Day 4: none

## APPLY/HOMEWORK

**Homework Assignment  (See also the assignments for Lessons 8.3 and 8.5.)**
____ Block Schedule:  Day 3: 9–38; Day 4: 39–53, 55–57, 61–71

**Reteaching the Lesson**
____ Practice Masters: CRB pages 55–57 (Level A, Level B, Level C)
____ Reteaching with Practice: CRB pages 58–59 or Practice Workbook with Examples
____ Personal Student Tutor

**Extending the Lesson**
____ Applications (Interdisciplinary): CRB page 61
____ Challenge: SE page 487; CRB page 62 or Internet

## ASSESSMENT OPTIONS

____ Checkpoint Exercises:   Day 3: TE page 481 or Transp.; Day 4: TE page 482 or Transp.
____ Daily Homework Quiz (8.4): TE page 487, CRB page 65, or Transparencies
____ Standardized Test Practice: SE page 487; TE page 487; STP Workbook; Transparencies

Notes _____

_____

_____

| CHAPTER PACING GUIDE | |
| --- | --- |
| **Day** | **Lesson** |
| 1 | Assess Ch. 7: 8.1 (all) |
| 2 | 8.2 (all); 8.3 (begin) |
| 3 | 8.3 (end); **8.4 (begin)** |
| 4 | **8.4 (end)**; 8.5 (begin) |
| 5 | 8.5 (end); 8.6 (begin) |
| 6 | 8.6 (end); 8.7 (begin) |
| 7 | 8.7 (end); Review Ch. 8 |
| 8 | Assess Ch. 8; 9.1 (all) |

NAME _____ DATE _____

# WARM-UP EXERCISES

For use before Lesson 8.4, pages 480–487

In the diagram, △**ABC** ~ △**DEF**.
Find the measure.

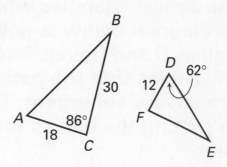

1. $m\angle F$

2. $m\angle A$

3. $m\angle B$

4. $FE$

........................................................

# DAILY HOMEWORK QUIZ

For use after Lesson 8.3, pages 472–479

## Are the polygons similar? If so, write a similarity statement.

1.

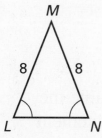

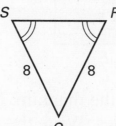

2.

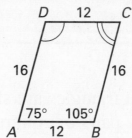

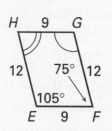

In Exercises 3–5, △**KLM** ~ △**PQR**.

3. Find the scale factor.

4. Find the length of $\overline{PQ}$.

5. Find the measure of $\angle Q$.

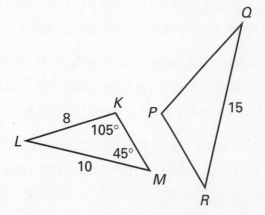

## *Application Lesson Opener*

For use with pages 480–487

**You can use similar triangles when playing miniature golf. In the diagram below, a golf ball putted from point *B* hits a wall at *D* and travels into the hole at *H*. By physics, the angles that the path of the ball makes with the wall are always congruent. The challenge for the golfer is in putting the ball so that it hits the wall at the correct point.**

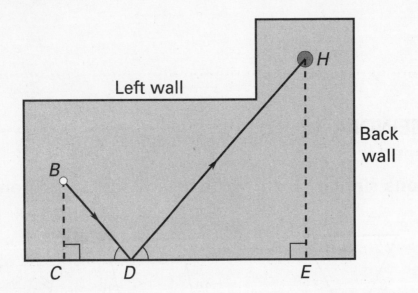

Left wall

Back
wall

B

H

C   D        E

1. Two similar right triangles are shown in the diagram. Name the triangles and any pairs of congruent angles. Write the statement of proportionality for the triangles.

2. Trace the diagram except for the path. Draw a path for the ball to reach the hole in one putt by hitting the *back* wall instead of the right wall. Use a protractor to verify congruent angles.

3. Trace the diagram except for the path. Draw a path for the ball to reach the hole in one putt by hitting *two* walls. (*Hint:* Hit the left wall first.) Use a protractor to verify congruent angles.

4. Design your own miniature golf hole that requires the ball to hit one or two walls. Draw at least one possible path to the hole.

# *Technology Activity*

**For use with pages 480–487**

**GOAL** **To construct similar triangles and to determine a relationship between the corresponding sides of similar triangles**

Similar figures are an important part of geometry. Many theorems can be proved by using the concept of similarity. In this activity, you will construct similar triangles and then determine a relationship between the corresponding sides of similar triangles.

## *Activity*

**1** Draw $\triangle ABC$.

**2** Draw $\triangle DEF$ similar to $\triangle ABC$ (use the dilation feature).

**3** Measure the angles in each triangle.

**4** Measure the sides in each triangle.

**5** Compare the ratio of the corresponding sides of the triangles.

## *Exercises*

**1.** What is true about the ratios of the corresponding sides in similar triangles?

**2.** For each of the following, determine whether $\triangle ABC$ is similar to $\triangle WXY$.

    **a.** $AB = 6, BC = 8, AC = 16$ and $WX = 3, XY = 4, WY = 8$

    **b.** $AB = 3, BC = 4, AC = 9$ and $WX = 1.5, XY = 2, WY = 4$

    **c.** $AB = 7, BC = 12, AC = 20$ and $WX = 21, XY = 35, WY = 60$

**3.** $\triangle COB$ is similar to $\triangle TIP$. Find the missing sides for each of the following.

    **a.** $CO = 5, TI = 15, CB = 11,$ and $IP = 23$

    **b.** $CB = 15, OB = 12, TP = 22.5,$ and $TI = 10.5$

# *Technology Activity Keystrokes*

**For use with pages 480–487**

## TI-92

1. Draw △*ABC*.

   [F3] 3 (Move cursor to location for *A*.) [ENTER] *A* (Move cursor to location

   for *B*.) [ENTER] *B* (Move cursor to location for *C*.) [ENTER] *C*

2. Draw △*DEF* similar to △*ABC*.

   (Plot a point not on △*ABC*.) [F2] 1

   (Move cursor to a blank area of the drawing page.) [F7] 6 [ENTER] 2

   [F5] 3 (Move cursor to △*ABC*.) [ENTER] (Move cursor to point not on △*ABC*.)

   [ENTER] (Move cursor to the number 2.) [ENTER]

3. Measure the angles in each triangle.

   [F6] 3 (Place cursor on one side of angle.) [ENTER] (Move cursor to vertex of

   desired angle.) [ENTER] (Place cursor on second side of angle.) [ENTER]

   Repeat this process for the other angles.

4. Measure the sides of each triangle.

   [F6] 1 (Place cursor on one endpoint of side.) [ENTER] (Move cursor to other

   endpoint of side.) [ENTER]

   Repeat this process for the other sides.

5. Calculate the ratios of the corresponding sides.

   [F6] 6 (Use cursor to highlight desired side.) [ENTER] [÷] (Highlight other side)

   [ENTER] [=] (The result will appear on the screen.)

## SKETCHPAD

1. Turn on the axes and the grid. Choose **Snap To Grid** from the **Graph** menu. Choose **Plot Points...** from the **Graph** menu. Enter the points $(1, -0.5)$ and $(1.5, -1.5)$, then click OK. Use the text tool to relabel the points $A$ and $B$, respectively. Draw $\overline{AB}$ using the segment straightedge tool.

2. Choose the point tool and plot a point not on $\triangle ABC$. Choose the translate selection arrow tool and select the point. Then choose **Mark Center** from the **Transform** menu. Select $\triangle ABC$ and choose **Dilate...** from the **Transform** menu. Enter 2 as the new scale factor, enter 1 as the old scale factor, and click OK. Use the text tool to label the new triangle $DEF$.

3. Measure the angles in each triangle. Choose the selection arrow tool, select the three points that make up the angle while holding down the shift key (the vertex must be the second point selected), and choose **Angle** from the **Measure** menu. Repeat for the other five angles.

4. Measure the sides of each triangle. Choose the selection arrow tool, select each side of the triangles while holding down the shift key, and choose **Length** from the **Measure** menu.

5. Calculate the ratios of the corresponding sides of the triangles. Choose **Calculate** from the **Measure** menu, click the measurement of one side of triangle $ABC$, click $\boxed{÷}$, click the corresponding side in triangle $FJO$, and click $\boxed{=}$. Repeat for the other two ratios.

## Keystrokes for Exercise 54 (p. 486)

### TI-92

1. Turn on the coordinate axes and the grid.

2. Draw line $k$ with a negative slope choosing points on the grid for the line.

    $\boxed{F2}$ 4 (Place cursor on a grid point.) $\boxed{ENTER}$ (Move cursor to a second grid point so that the line will have a negative slope.) $\boxed{ENTER}$ $k$

3. Draw a right triangle with the hypotenuse on line $k$ and the legs parallel to the $x$- and $y$-axes.

    $\boxed{F2}$ 5 (Choose a grid point that lies on line $k$.) $\boxed{ENTER}$ $A$ (Move cursor parallel to the $y$-axis to another grid point.) $\boxed{ENTER}$ $B$ (Place cursor on point $B$.) $\boxed{ENTER}$ (Move cursor parallel to the $x$-axis across line $k$ to a grid point.) $\boxed{ENTER}$ (Label the intersection of this segment and line $k$ point C.) $\boxed{F3}$ 3 $\boxed{ENTER}$ $C$

4. Draw a second right triangle with the hypotenuse on line $k$ and legs parallel to the $x$- and $y$-axes.

    $\boxed{F2}$ 5 (Choose a grid point that lies on line $k$.) $\boxed{ENTER}$ $D$ (Move cursor parallel to the $y$-axis to another grid point.) $\boxed{ENTER}$ $E$ (Place cursor on point $E$.) $\boxed{ENTER}$

## LESSON
## 8.4
**CONTINUED**

NAME _____ DATE _____

# *Technology Activity Keystrokes*

**For use with page 486**

(Move cursor parallel to the *x*-axis across line *k* to a grid point.) `ENTER` (Label the intersection of this segment and line *k* point *F*.) `F3` 3 (Place cursor on intersection point.) `ENTER` *F*

5. Find the lengths of $\overline{AB}, \overline{BC}, \overline{DE},$ and $\overline{EF}$.

`F6` 1 (Move cursor to $\overline{AB}$.) `ENTER` (Move cursor to *B*.) `ENTER` (Move cursor to *C*.) `ENTER` (Move cursor to $\overline{DE}$.) `ENTER` (Move cursor to *E*.) `ENTER` (Move cursor to *F*.) `ENTER`

6. Calculate the slope.

`F6` 6 (Use cursor to highlight length of $\overline{AB}$.) `ENTER` `÷` (Use cursor to highlight length of $\overline{BC}$.) `ENTER` `ENTER` (The result will appear on the screen.)

`F6` 6 (Use cursor to highlight length of $\overline{DE}$.) `ENTER` `÷` (Use cursor to highlight length of $\overline{EF}$.) `ENTER` `ENTER` (The result will appear on the screen.)

## Keystrokes for Exercise 54

### SKETCHPAD

1. Turn on the coordinate axes and the grid. Choose **Snap to Grid** from the **Graph** Menu.

2. Use the line straightedge tool to draw line *k* with a negative slope choosing two points on the grid for the line.

3. Draw a right triangle with the hypotenuse on line *k* and one leg parallel to the *x*-axis and the other leg parallel to the *y*-axis using the segment straight-edge tool.

4. Draw a second right triangle using the same restrictions in Step 3.

5. Find the lengths of the legs of both right triangles. Select the segments using the selection arrow tool. Then choose **Length** from the **Measure** menu.

6. Calculate the slope. Choose **Calculate** from the **Measure** menu. Click on the measure of a two leg of the first triangle, click on the division sign, click on the measure of the first triangle's other leg, and click OK. Repeat this process for the second triangle.

NAME _____ DATE _____

# Practice A

For use with pages 480–487

**Which triangles are similar to △EFG? Explain.**

**1.**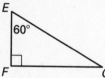

**A.**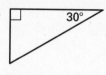

**B.**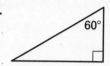

**C.**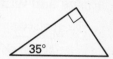

**2.**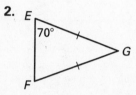

**A.**

**B.**

**C.**

**Decide whether △ABC and △DEF are similar, not similar, or cannot be determined for the given information.**

**3.**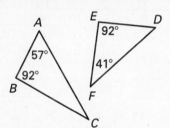

**4.**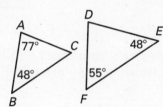

**5.**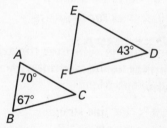

**The triangles shown are similar. List all the pairs of congruent angles and write the statement of proportionality.**

**6.**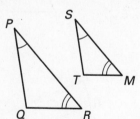

**7.**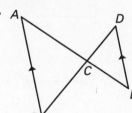

**8.**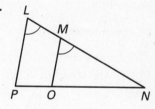

**Use the diagram to complete the following.**

**9.** $\triangle MON \sim$ ___?___

**10.** $\dfrac{MN}{?} = \dfrac{ON}{?} = \dfrac{MO}{?}$

**11.** $\dfrac{16}{?} = \dfrac{?}{10}$

**12.** $\dfrac{?}{16} = \dfrac{8}{?}$

**13.** Solve for $x$ and $y$.

Lesson 8.4

## Practice B

For use with pages 480–487

**The triangles shown are similar. List all the pairs of congruent angles and write the statement of proportionality.**

1.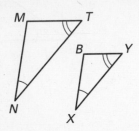

2.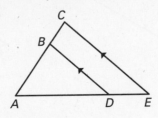

3.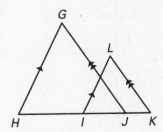

**Use the diagram to complete the following.**

4. $\triangle TIR \sim \triangle$ ___?___

5. $\dfrac{TI}{?} = \dfrac{IR}{?} = \dfrac{RT}{?}$

6. $\dfrac{24}{?} = \dfrac{?}{10}$

7. $\dfrac{?}{24} = \dfrac{12}{?}$

8. Solve for $x$ and $y$.

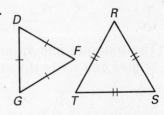

**Determine whether the triangles can be proved similar. If they are similar, write a similarity statement. If they are not similar, explain why.**

9.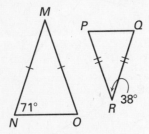

10.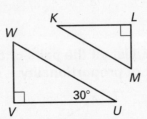

11.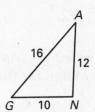

12.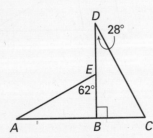

13.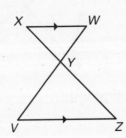

14.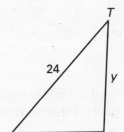

**Write a paragraph or a two-column proof.**

15. **Given:** $\overline{DE}$ is a midsegment of $\triangle ABC$.
**Prove:** $\triangle ABC \sim \triangle DBE$

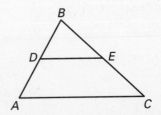

Lesson 8.4

NAME _____ DATE _____

# Practice C

For use with pages 480–487

**The triangles shown are similar. List all the pairs of congruent angles and write the statement of proportionality.**

**1.**
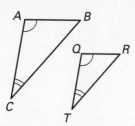

**2.**
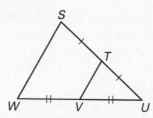

**3.**
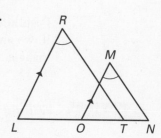

**Determine whether the triangles can be proved similar. If they are similar, write a similarity statement. If they are not similar, explain why.**

**4.**
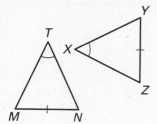

**5.**
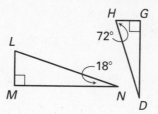

**6.**
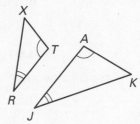

**7.**
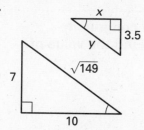

**8.**
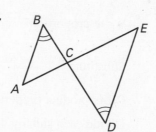

**9.**
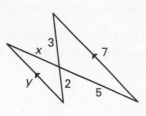

**The triangles are similar. Find the value of the variables.**

**10.**
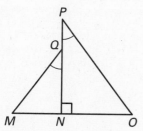

**11.**
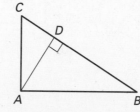

**Write a paragraph or a two-column proof.**

**12. Given:** $\triangle ABC$ is a right triangle.

      $\overline{AD}$ is an altitude.

  **Prove:** $\triangle ABC \sim \triangle DAC$

NAME _____ DATE _____

# Reteaching with Practice

For use with pages 480–487

**GOAL** **Identify similar triangles**

> ## VOCABULARY
>
> **Postulate 25   Angle-Angle (AA) Similarity Postulate**
> If two angles of one triangle are congruent to two angles of another
> triangle, then the two triangles are similar.

**EXAMPLE 1** *Writing Proportionality Statements*

In the diagram $\triangle ABC \sim \triangle DEC$.

    **a.** Write the statement of proportionality.

    **b.** Find $m\angle D$.

    **c.** Find the length of $\overline{CE}$.

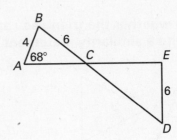

**SOLUTION**

**a.** $\dfrac{AB}{DE} = \dfrac{BC}{EC} = \dfrac{CA}{CD}$

**b.** $\angle A \cong \angle D$, so $m\angle D = 68°$.

**c.**

| | |
|---|---|
| $\dfrac{AB}{DE} = \dfrac{BC}{EC}$ | Write proportion. |
| $\dfrac{4}{6} = \dfrac{6}{CE}$ | Substitute. |
| $4 \cdot CE = 36$ | Cross product property |
| $CE = 9$ | Divide each side by 4. |

So, the length of $\overline{CE}$ is 9.

## Exercises for Example 1

**The triangles shown are similar. List all the pairs of congruent
angles and write the statement of proportionality. Find the value of
*x*.**

**1.**

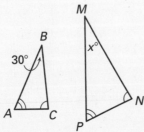

**2.**

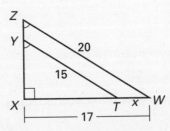

**3.**

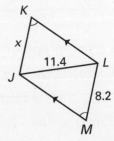

Lesson 8.4

NAME _____ DATE _____

# Reteaching with Practice

For use with pages 480–487

**EXAMPLE 2** *Proving that Two Triangles are Similar*

Determine whether the triangles can be proved
similar. If they are similar, write a similarity
statement. If they are not similar, explain why.

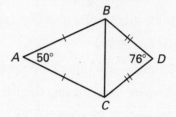

### SOLUTION

In △*ABC*, you are given that *m*∠*A* = 50°. Because △*ABC* is an isosceles trian-
gle, you know that $m\angle ABC = m\angle ACB = \dfrac{180° - 50°}{2} = \dfrac{130°}{2} = 65°$. Similarly,
you can find the angles of △*DBC* to be 76°, 52°, and 52°. Because the angles in
△*ABC* are not congruent to the angles in △*DBC*, the triangles are not similar.

*Exercises for Example 2*
.................................................................................................................

**Determine whether the triangles can be proved similar. If they are
similar, write a similarity statement. If they are not similar, explain
why.**

**4.**

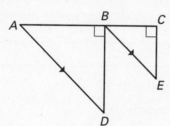

**5.**

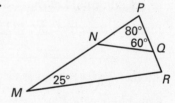

**6.**

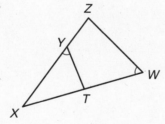

NAME _____ DATE _____

# Quick Catch-Up for Absent Students

**For use with pages 480–487**

The items checked below were covered in class on (date missed) _____

**Lesson 8.4: Similar Triangles**

____ **Goal 1:** Identify similar triangles. (pp. 480–481)

*Material Covered:*

    ____ Example 1: Writing Proportionality Statements

    ____ Activity: Investigating Similar Triangles

    ____ Example 2: Proving that Two Triangles are Similar

    ____ Student Help: Look Back

    ____ Example 3: Why a Line Has Only One Slope

____ **Goal 2:** Use similar triangles in real-life problems. (p. 482)

*Material Covered:*

    ____ Example 4: Using Similar Triangles

    ____ Example 5: Using Scale Factors

____ Other (specify) _____

_____

**Homework and Additional Learning Support**

    ____ Textbook (specify) _pp. 483–487_____

_____

    ____ Internet: Extra Examples at www.mcdougallittell.com

    ____ *Reteaching with Practice* worksheet (specify exercises)_____

    ____ *Personal Student Tutor* for Lesson 8.4

NAME _____ DATE _____

# *Interdisciplinary Application*

**For use with pages 480–487**

## Theodolite

**CIVIL ENGINEERING**   A theodolite is an instrument used by surveyors to measure distances indirectly based on the measures of similar right triangles. It consists of a telescope on a plate marking 360 degrees around and a vertical wheel to mark angles of elevation. It is mounted on a tripod, allowing free rotation about its horizontal and vertical axes. With this instrument, surveyors are able to construct triangles with known angles and lengths using landmarks like trees, large rocks, or stakes driven into the ground. The diagram below shows an example of a triangle constructed with two poles and a tree across a river.

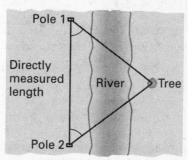

Refined versions of this basic surveying tool have included a theolodite with a horizontal line across the middle and another horizontal line a little above this center line. If a theolodite were 12 inches long and the horizontal lines were $\frac{1}{4}$ inch apart, then a right triangle would be formed as shown in the diagrams below. A rod with measurements marked on it is sighted through the theodolite. The distance to the stick is related to to the measurements read between the two horizontal lines.

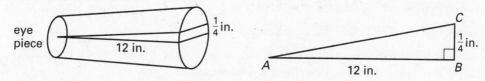

## In Exercises 1–5, use the diagrams above and the following information.

A surveyor is using the theodolite in the diagram above and the measure on the rod that he reads between the lines in the scope is 14 inches.

1. Make a triangle diagram representing this situation using the measurements for the theolodite and the measurements on the rod. Point $A$ is the eyepiece and points $B$ and $C$ are the top and bottom lines in the scope (see diagram above). Points $D$ and $E$ are the top and bottom marks on the 14 inch rod.

2. Which side of $\triangle ABC$ corresponds to the 14 inch measurement?

3. Which side of $\triangle ABC$ corresponds to the distance to the rod?

4. How far is the rod from the surveyor? Answer in feet.

5. How far away would the rod have to be from the surveyor to see the whole top half of a 6 foot rod between the two lines? Answer in feet.

NAME _____ DATE _____

# Challenge: Skills and Applications

For use with pages 480–487

1. Refer to the diagram, where $\overline{VW} \parallel \overline{YZ}$.

   a. Write a similarity statement.

   b. Write a paragraph proof for your result.

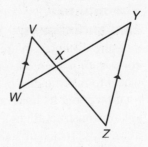

2. In the diagram, $ABCD$ is a parallelogram.

   a. Name three triangles that are similar to $\triangle BEF$.
   (For each triangle, give the vertices in the correct, corresponding order.)

   b. Write a paragraph proof for your result.

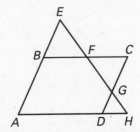

**In Exercises 3–8, refer to the diagram. Find the coordinates of the missing point so that the similarity statement is true. (There may be more than one correct answer.)**

3. Given $\triangle PQR \sim \triangle STU$, find the coordinates of $U$.

4. Given $\triangle PQR \sim \triangle VST$, find the coordinates of $V$.

5. Given $\triangle PQR \sim \triangle SWT$, find the coordinates of $W$.

6. Given $\triangle PQR \sim \triangle TSX$, find the coordinates of $X$.

7. Given $\triangle PQR \sim \triangle YTS$, find the coordinates of $Y$.

8. Given $\triangle PQR \sim \triangle TZS$, find the coordinates of $Z$.

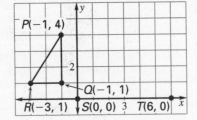

9. Determine if the following conjecture is true or false. If it is true, write a paragraph proof; if it is false, sketch or describe a counterexample.

   If the corresponding angles of quadrilaterals $ABCD$ and $EFGH$ are congruent, then $ABCD \sim EFGH$.

10. To estimate the radius of the sun, a student punches a tiny hole in a piece of paper and allows the sun to shine through the hole, forming an image on a screen 200 cm away. If the image has a radius of 0.6 cm and the student knows that the sun is 150,000,000 km away, what is the student's estimate of the radius of the sun? (Illustration is not to scale.)

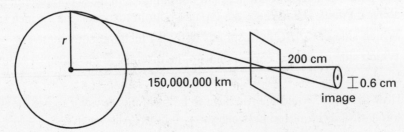

Lesson 8.4

TEACHER'S NAME _____ CLASS _____ ROOM _____ DATE _____

## Lesson Plan

2-day lesson (See *Pacing the Chapter*, TE pages 454C–454D)          For use with pages 488–496

**GOALS**   1. **Use similarity theorems to prove that two triangles are similar.**
2. **Use similar triangles to solve real-life problems.**

State/Local Objectives _____

_____

## ✓ Check the items you wish to use for this lesson.

### STARTING OPTIONS
____ Homework Check: TE page 483: Answer Transparencies
____ Warm-Up or Daily Homework Quiz: TE pages 488 and 487, CRB page 65, or Transparencies

### TEACHING OPTIONS
____ Lesson Opener (Activity): CRB page 66 or Transparencies
____ Technology Activity with Keystrokes: CRB pages 67–69
____ Examples:   Day 1: 1–4, SE pages 488–490; Day 2: 5–6, SE pages 490–491
____ Extra Examples:   Day 1: TE pages 489–490 or Transp.; Day 2: TE page 491 or Transp.
____ Closure Question: TE page 491
____ Guided Practice: SE page 492   Day 1: Exs. 1–5; Day 2: none

### APPLY/HOMEWORK
#### Homework Assignment
____ Basic   Day 1: 6–18 even, 19–28, 30; Day 2: 7–17 odd, 31–37, 39–47; Quiz 2: 1–7
____ Average   Day 1: 6–18 even, 19–28, 30; Day 2: 7–17 odd, 29, 31–37, 39–47; Quiz 2: 1–7
____ Advanced   Day 1: 6–18 even, 19–28, 30; Day 2: 7–17 odd, 29, 31–38, 39–47; Quiz 2: 1–7

#### Reteaching the Lesson
____ Practice Masters: CRB pages 70–72 (Level A, Level B, Level C)
____ Reteaching with Practice: CRB pages 73–74 or Practice Workbook with Examples
____ Personal Student Tutor

#### Extending the Lesson
____ Cooperative Learning Activity: CRB page 76
____ Applications (Real-Life): CRB page 77
____ Math & History: SE page 496; CRB page 78; Internet
____ Challenge: SE page 495; CRB page 79 or Internet

### ASSESSMENT OPTIONS
____ Checkpoint Exercises:   Day 1: TE pages 489–490 or Transp.; Day 2: TE page 491 or Transp.
____ Daily Homework Quiz (8.5): TE page 495, CRB page 83, or Transparencies
____ Standardized Test Practice: SE page 495; TE page 495; STP Workbook; Transparencies
____ Quiz (8.4–8.5): SE page 496; CRB page 80

Notes _____

_____

_____

TEACHER'S NAME _____ CLASS _____ ROOM _____ DATE _____

# Lesson Plan for Block Scheduling

1-day lesson (See *Pacing the Chapter*, TE pages 454C–454D)          For use with pages 488–496

**GOALS**
1. Use similarity theorems to prove that two triangles are similar.
2. Use similar triangles to solve real-life problems.

State/Local Objectives _____

_____

_____

✓ **Check the items you wish to use for this lesson.**

## STARTING OPTIONS

____ Homework Check: TE page 483: Answer Transparencies
____ Warm-Up or Daily Homework Quiz: TE pages 488 and 487, CRB page 65, or Transparencies

## TEACHING OPTIONS

____ Lesson Opener (Activity): CRB page 66 or Transparencies
____ Technology Activity with Keystrokes: CRB pages 67–69
____ Examples:   Day 4: 1–4, SE pages 488–490; Day 5: 5–6, SE pages 490–491
____ Extra Examples:   Day 4: TE pages 489–490 or Transp.; Day 5: TE page 491 or Transp.
____ Closure Question: TE page 491
____ Guided Practice: SE page 492   Day 4: Exs. 1–5; Day 5: none

## APPLY/HOMEWORK

**Homework Assignment (See also the assignments for Lessons 8.4 and 8.6.)**
____ Block Schedule:  Day 4: 6–18 even, 19–28, 30; Day 5: 7–17 odd, 29, 31–37, 39–47; Quiz 2: 1–7

## Reteaching the Lesson

____ Practice Masters: CRB pages 70–72 (Level A, Level B, Level C)
____ Reteaching with Practice: CRB pages 73–74 or Practice Workbook with Examples
____ Personal Student Tutor

## Extending the Lesson

____ Cooperative Learning Activity: CRB page 76
____ Applications (Real-Life): CRB page 77
____ Math & History: SE page 496; CRB page 78; Internet
____ Challenge: SE page 495; CRB page 79 or Internet

## ASSESSMENT OPTIONS

____ Checkpoint Exercises:   Day 4: TE pages 489–490 or Transp.; Day 5: TE page 491 or Transp.
____ Daily Homework Quiz (8.5): TE page 495, CRB page 83, or Transparencies
____ Standardized Test Practice: SE page 495; TE page 495; STP Workbook; Transparencies
____ Quiz (8.4–8.5): SE page 496; CRB page 80

| **CHAPTER PACING GUIDE** | |
|:---:|:---|
| **Day** | **Lesson** |
| 1 | Assess Ch. 7: 8.1 (all) |
| 2 | 8.2 (all); 8.3 (begin) |
| 3 | 8.3 (end); 8.4 (begin) |
| 4 | 8.4 (end); **8.5 (begin)** |
| 5 | **8.5 (end)**; 8.6 (begin) |
| 6 | 8.6 (end); 8.7 (begin) |
| 7 | 8.7 (end); Review Ch. 8 |
| 8 | Assess Ch. 8; 9.1 (all) |

Notes _____

_____

_____

## LESSON 8.5

NAME _____ DATE _____

# WARM-UP EXERCISES

For use before Lesson 8.5, pages 488–496

**In the figure below, find a pair of similar triangles and use them to complete Exercises 1–3.**

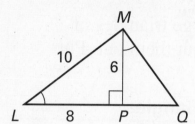

**1.** Write the statement of similarity for the two triangles.

**2.** Explain why the two triangles are similar.

**3.** Find $MQ$.

............................................................................................................

# DAILY HOMEWORK QUIZ

For use after Lesson 8.4, pages 480–487

**Use the diagram for Exercises 1–5.**

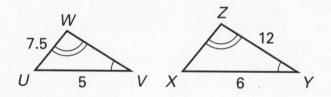

**1.** $\triangle UVW \sim \triangle$ _____

**2.** What is the scale factor of $\triangle UVW$ to $\triangle XYZ$?

**3.** What is $VW$?

**4.** What is $XZ$?

**5.** If $m\angle U = 50°$ and $m\angle Y = 30°$, what is $m\angle Z$?

Lesson 8.5

NAME ———————————————————— DATE ————

## *Activity Lesson Opener*

For use with pages 488–496

**SET UP:** Work in a group.
**YOU WILL NEED:** • scissors • construction paper
• ruler • protractor

1. Each member of the group should draw two large triangles of any type on a sheet of construction paper and cut them out. Put all the triangles for your group in a pile.

**In Exercises 2–4, you will be creating similar triangles.
Do not use a protractor or ruler to create the triangles.**

2. Choose a triangle from the pile and create a similar triangle as follows: locate the midpoints of two sides of the triangle by paper folding. Place the triangle on a new sheet of paper and trace the angle included by the midpoints. Trace the sides up to the midpoints. Remove the triangle and draw a segment for the third side of a new triangle. Cut out the triangle. Two pairs of corresponding sides are proportional, with the included angles congruent, so the two triangles are similar by the Side-Angle-Side (SAS) Similarity Theorem. Use a protractor and a ruler to verify similarity.

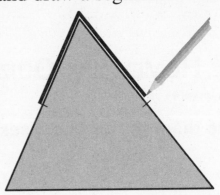

3. Choose another triangle from the pile and create a similar triangle as follows: first create a congruent triangle by tracing and cutting. Then cut a narrow strip of the same width off each side of the new triangle, parallel to the side. Corresponding sides are proportional, so the two triangles are similar by the Side-Side-Side (SSS) Similarity Theorem. Use a protractor and a ruler to verify similarity.

4. Choose another triangle from the pile and create a similar triangle by using the AA Similarity Postulate. As a group, create some triangles that are larger and some that are smaller. Share your methods as you work. Use a protractor and a ruler to verify similarity.

NAME _____ DATE _____

# Technology Activity

For use with pages 488–496

**GOAL** **To determine if two triangles are similar given a special condition**

If a line parallel to one side of a triangle intersects the other two sides, will the resulting smaller triangle be similar to the original triangle? Perform the following activity and answer the Exercises to find out.

## Activity

❶ Draw △ABC and point D on $\overline{AB}$.

❷ Draw $\overline{ED}$ parallel to $\overline{BC}$ (see figure).

❸ Measure the sides of △ABC.

❹ Measure the sides of △ADE.

❺ Compare the ratios of the corresponding sides.

❻ Drag a vertex of △ABC and notice the results.

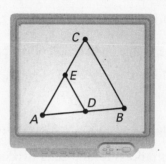

## Exercises

1. When dragging a vertex of △ABC, did the ratios of the corresponding sides remain the same?

2. The activity's construction illustrated which method of proving triangles similar?

3. A parallel line was used in the activity's construction. What special angle pairs could be shown congruent?

4. Using the information from Exercise 3, which method could be used to prove the triangles similar?

# Technology Activity Keystrokes

For use with pages 488–496

## TI-92

1. Construct $\triangle ABC$ using the triangle command. ( F3 3)

   Place point $D$ on $\overline{AB}$.

   F2 2 (Move cursor to $\overline{AB}$.) ENTER $D$

2. Draw a line parallel to $\overline{BC}$ through point $D$.

   F4 2 (Move cursor to $D$.) ENTER (Move cursor to $\overline{BC}$.) ENTER

   Plot point $E$ as the intersection of the parallel line and side $AC$.

   F2 3 (Move to intersection of parallel line and $\overline{AC}$.) ENTER $E$

3. Measure the sides of $\triangle ABC$.

   F6 1 (Place cursor on $A$.) ENTER (Move cursor to $B$.) ENTER

   Repeat this process for the other two sides.

4. Measure the sides of $\triangle ADE$.

   F6 1 (Place cursor on point $A$.) ENTER (Move cursor to point $D$.) ENTER

   Repeat this process for the other two sides.

5. Compare the ratios of the corresponding sides: $\dfrac{AB}{AD}$, $\dfrac{AC}{AE}$, and $\dfrac{BC}{DE}$.

   F6 6 (Move cursor to highlight length of $\overline{AB}$.) ENTER ÷ (Move cursor

   to highlight length of $\overline{AD}$.) ENTER ENTER (The result will appear on the screen.)

   Repeat this process for the other two ratios.

6. Drag a vertex of $\triangle ABC$.

   F1 1 (Place cursor on $A$.) ENTER (Use the drag key 🖐 and

   the cursor pad to drag the vertex.)

**Geometry**
Chapter 8 Resource Book

LESSON
**8.5**
CONTINUED

NAME _____ DATE _____

# *Technology Activity Keystrokes*

**For use with pages 488–496**

*Lesson 8.5*

## SKETCHPAD

1. Draw $\triangle ABC$ using the segment straightedge tool. Plot point $D$ on $\overline{AB}$ using the point tool.

2. Draw a line parallel to $\overline{BC}$ through D. Use the selection arrow tool to select $D$, hold down the shift key, select $\overline{BC}$, and select **Parallel Line** from the **Construct** menu. Select the parallel line and $\overline{AC}$. Choose **Point At Intersection** from the **Construct** menu.

3. Measure the sides of $\triangle ABC$. Using the selection arrow tool, select a point, hold down the shift key and select another point that makes up the side, and choose **Distance** from the **Measure** menu. Repeat this process for the other sides.

4. Measure the sides of $\triangle ADE$. See Step 3.

5. Compare the ratios of the corresponding sides: $\dfrac{AB}{AD}$, $\dfrac{AC}{AE}$, and $\dfrac{BC}{DE}$.

   Choose **Calculate** from the **Measure** menu, click the length of $\overline{AB}$, click $\boxed{\div}$, click the length of $\overline{AD}$, and click OK. Repeat this process for the other two ratios.

6. Using the translate selection arrow tool, drag a vertex of $\triangle ABC$.

NAME _____ DATE _____

## Practice A

For use with pages 488–496

**Name a postulate or theorem that can be used to prove that the two triangles are similar. Then, write a similarity statement.**

1.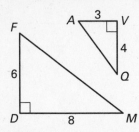

2.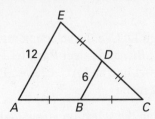

3.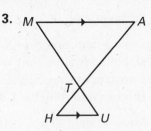

**Determine which two of the three given triangles are similar. Find the scale factor for the pair.**

4.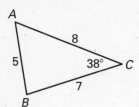

5.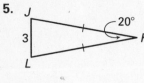

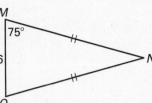

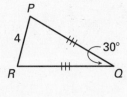

**Are the triangles similar? If so, state the similarity and the postulate or theorem that justifies your answer.**

6.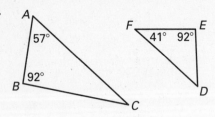

7.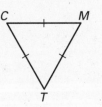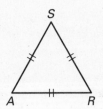

**Decide whether the statement is *true* or *false*. Explain your reasoning.**

8. If an acute angle of a right triangle is congruent to an acute angle of another right triangle, then the triangles are similar.

9. All equilateral triangles are similar.

10. If two triangles are congruent, then they are similar.

11. If two triangles are similar, then they are congruent.

12. All isosceles triangles with a 40° vertex angle are similar.

NAME _____ DATE _____

## Practice B

For use with pages 488–496

**Name a postulate or theorem that can be used to prove that the two
triangles are similar. Then, write a similarity statement.**

**1.**
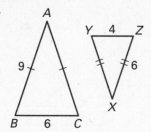

**2.**
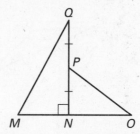

**3.**
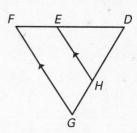

**Are the triangles similar? If so, state the similarity and the postulate
or theorem that justifies your answer.**

**4.**
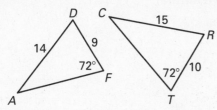

**5.**
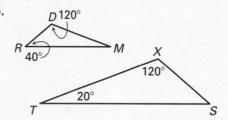

**Draw the given triangles roughly to scale. Then, name a postulate or
theorem that can be used to prove that the triangles are similar.**

**6.** The side lengths of $\triangle ABC$ are 3, 4, and 6, and the side lengths of $\triangle XYZ$ are
6, 8, and 12.

**7.** In $\triangle ABC$, $m\angle A = 15°$ and $m\angle B = 80°$. In $\triangle XYZ$, $m\angle Y = 80°$ and
$m\angle Z = 85°$.

**8.** In $\triangle ABC$, $m\angle B = 60°$, $AB = 6$, and $BC = 12$. In $\triangle XYZ$, $m\angle Y =
60°$, $XY = 3$, and $YZ = 6$.

**Use the diagram shown to complete the statements.**

**9.** $\triangle AEB \sim$ ____?____

**10.** $m\angle DEC =$ ____?____

**11.** $m\angle EBA =$ ____?____

**12.** $EC =$ ____?____

**13.** perimeter $\triangle DEC$: perimeter $\triangle BEA =$ ____?____

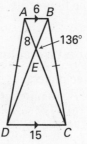

**In Exercises 14 and 15, use the diagram at the right.**

To determine the height of a very tall pine tree, you place a mirror on
the ground and stand where you can see the top of the tree, as shown.

**14.** How tall is the tree?

**15.** Your little sister wants to see the top of the tree also. However, she
is only 4 feet tall. Leaving the mirror 2 feet from her feet, how far
from the base of the tree should the mirror be placed?

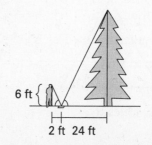

**Geometry**
Chapter 8 Resource Book

## Practice C

For use with pages 488–496

**Are the triangles similar? If so, state the similarity and the postulate or theorem that justifies your answer.**

1.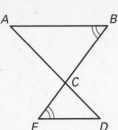

2.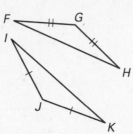

3.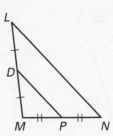

**Draw the given triangles roughly to scale. Then, name a postulate or theorem that can be used to prove that the triangles are similar.**

4. In $\triangle ABC$, $m\angle A = 38°$ and $m\angle B = 94°$. In $\triangle XYZ$, $m\angle Y = 94°$ and $m\angle Z = 48°$.

5. The ratio of $AB$ to $XY$ is 2:3. In $\triangle ABC$, $m\angle B = 75°$, and in $\triangle XYZ$, $m\angle Y = 75°$. The ratio of $BC$ to $YZ$ is 2:3.

6. In $\triangle ABC$, $m\angle B = 50°$, $AB = 4$, and $BC = 9$. In $\triangle XYZ$, $m\angle Y = 50°$, $XY = 2$ and $YZ = 4.5$.

**Use the diagram shown to complete the statements.**

7. $m\angle DGE =$ ___?___

8. $m\angle EDG =$ ___?___

9. $FD =$ ___?___

10. $GD =$ ___?___

11. $EG =$ ___?___

12. Name the three pairs of triangles that are similar in the figure.

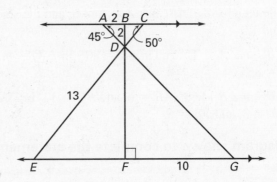

**Write a paragraph or a two-column proof.**

13. **Given:** $\triangle ABC$ is equilateral.
   $\overline{DE}, \overline{DF}, \overline{EF}$ are midsegments.
   **Prove:** $\triangle ABC \sim \triangle FED$

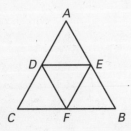

14. **Given:** $ABCD$ is a trapezoid
   with $\overline{AD}$ and $\overline{BC}$ as bases.
   **Prove:** $\triangle EAD \sim \triangle EBC$

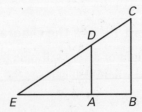

**Geometry**
Chapter 8 Resource Book

NAME _____ DATE _____

# *Reteaching with Practice*

For use with pages 488–496

**GOAL** **Use similarity theorems to prove that two triangles are similar**

---

### VOCABULARY

**Theorem 8.2   Side-Side-Side (SSS) Similarity Theorem**
If the corresponding sides of two triangles are proportional, then the triangles are similar.

**Theorem 8.3   Side-Angle-Side (SAS) Similarity Theorem**
If an angle of one triangle is congruent to an angle of a second triangle and the lengths of the sides including these angles are proportional, then the triangles are similar.

---

**EXAMPLE 1** *Using the SSS Similarity Theorem*

Which of the following triangles are similar?

 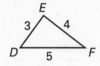 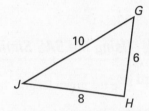

### SOLUTION

To decide which, if any, of the triangles are similar, you need to consider the ratios of the lengths of corresponding sides.

*Ratios of Side Lengths of △ABC and △DEF*

$$\frac{BC}{DE} = \frac{1}{3}$$

Shortest sides

$$\frac{CA}{DF} = \frac{3}{5}$$

Longest sides

$$\frac{AB}{EF} = \frac{2}{4} = \frac{1}{2}$$

Remaining sides

Because the ratios are not equal, △ABC and △DEF are not similar.

*Ratios of Side Lengths of △GHJ and △DEF*

$$\frac{GH}{DE} = \frac{6}{3} = \frac{2}{1}$$

Shortest sides

$$\frac{GJ}{DF} = \frac{10}{5} = \frac{2}{1}$$

Longest sides

$$\frac{HJ}{EF} = \frac{8}{4} = \frac{2}{1}$$

Remaining sides

Because the ratios are equal, △GHJ ~ △DEF.

Since △DEF is similar to △GHJ and △DEF is not similar to △ABC, △GHJ is not similar to △ABC.

---

NAME _____     DATE _____

# *Reteaching with Practice*

For use with pages 488–496

## Exercises for Example 1

**Determine which two of the three given triangles are similar.**

**1.**
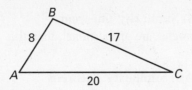
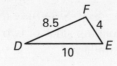
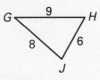

**2.**
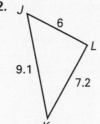
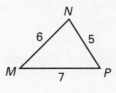
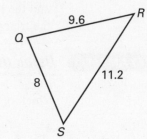

---

**EXAMPLE 2**   *Using the SAS Similarity Theorem*

Use the given lengths to prove that △ABC ~ △DEC.

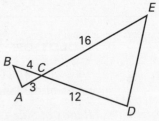

### SOLUTION

Begin by finding the ratios of the lengths of the corresponding sides.

$$\frac{AC}{DC} = \frac{3}{12} = \frac{1}{4} \qquad\qquad \frac{BC}{EC} = \frac{4}{16} = \frac{1}{4}$$

So, the side lengths AC and BC of △ABC are proportional to the corresponding side lengths DC and EC of △DEC. The included angle in △ABC is ∠BCA; the included angle in △DEC is ∠ECD. Because these two angles are vertical angles, they are congruent. So, by the SAS Similarity Theorem, △ABC ~ △DEC.

## Exercises for Example 2

**Prove that the two triangles are similar.**

**3.**
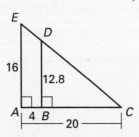

**4.**
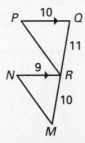

**5.**
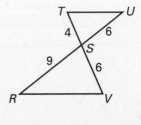

# *Quick Catch-Up for Absent Students*

**For use with pages 488–496**

The items checked below were covered in class on (date missed) _____

**Lesson 8.5: Proving Triangles are Similar**

_____ **Goal 1:** Use similarity theorems to prove that two triangles are similar. (pp. 488–489)

*Material Covered:*

_____ Example 1: Proof of Theorem 8.2

_____ Student Help: Study Tip

_____ Example 2: Using the SSS Similarity Theorem

_____ Example 3: Using the SAS Similarity Theorem

_____ **Goal 2:** Use similar triangles to solve real-life problems. (pp. 490–491)

*Material Covered:*

_____ Example 4: Using a Pantograph

_____ Example 5: Finding Distance Indirectly

_____ Example 6: Finding Distance Indirectly

_____ Other (specify) _____

_____

**Homework and Additional Learning Support**

_____ Textbook (specify) pp. 492–496 _____

_____

_____ Internet: Extra Examples at www.mcdougallittell.com

_____ *Reteaching with Practice* worksheet (specify exercises)_____

_____ *Personal Student Tutor* for Lesson 8.5

NAME _____ DATE _____

# Cooperative Learning Activity

For use with pages 488–496

**GOAL** **To use similar triangles to estimate the height of objects that would be difficult to measure directly**

**Materials:** measuring tape or meter stick, mirror with cross hairs in center, paper and pencil to record data, masking tape

## Exploring Similar Triangles

Similar triangles can be used to find distances that are difficult to measure directly. Triangles that are similar have sides that are proportional in length. Similar triangles can be used to solve for an unknown distance. One technique measures distance using shadows. Another method takes advantage of the reflective properties of mirrors and similarity to find an unknown distance.

## Instructions

**1** Locate a tall object that would be difficult to measure directly (flagpole, tree, school building, etc).

**2** Measure the height of one member of the group. Have this observer stand so that the tip of his or her shadow coincides with the tip of the object's shadow.

**3** Measure the distance from the observer to the object and the distance from the observer to the tip of the shadow.

**4** Calculate the height of the object.

**5** Repeat Steps 1–4 to calculate the height of another tall object.

**6** Place the mirror flat on the ground between the observer and the object measured in Step 1. Move the observer back and forward until the observer sees the top of the object over the cross hairs of the mirror (the cross hairs are created by making two lines with tape intersecting at right angles in the center of the mirror).

**7** Measure the distance from the observer to the mirror and the distance from the mirror to the base of the object.

**8** Calculate the height of the object.

**9** Repeat Steps 6–8 to calculate the height of the object measured in Step 5.

## Analyzing the Results

**1.** Are the calculated heights of the objects the same using each method? If not, which method do you think is most accurate?

**Geometry**
Chapter 8 Resource Book

NAME _____ DATE _____

# Real-Life Application: When Will I Ever Use This?

**For use with pages 488–496**

## Backyard Basketball

You want to put up a basketball hoop in your backyard. Regulation height for the basketball hoop is 10 feet or 120 inches. After a couple of attempts at guessing the correct height and checking, you decide to use shadows to estimate the height. You are 68 inches tall, and your shadow is 62 inches long. You are standing 41 inches from the pole and the tip of your shadow coincides with the tip of the basketball hoop's shadow as shown in the diagram below.

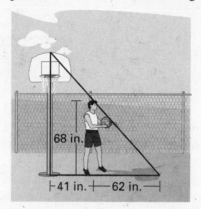

1. From the illustration, draw a diagram that describes the situation and label the given lengths of the corresponding sides using an *x* for the height of the basketball hoop.

2. Are the triangles similar? If so, state the postulate or theorem that justifies your answer.

3. How high did you hang the goal? Round to the nearest inch.

4. Is the basketball hoop currently at regulation height?

5. You put a stick in the ground to mark where the shadow of the basketball hoop should be and adjust the height of the hoop until the rim's shadow is on the stick. How far away from the base of the pole should you put the stick? Round to the nearest inch.

NAME _____ DATE _____

# *Math and History Application*

For use with page 496

**HISTORY**   Thousands of years ago, Greek mathematicians became interested in the *golden ratio*, a ratio of about 1:1.618. Psychologists have determined that rectangles whose side lengths are in the golden ratio are especially pleasing to look at. The ancient Greeks based much of their art and architecture on the golden ratio, including the Parthenon in Athens (Figure 1).

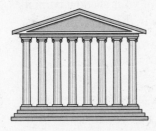
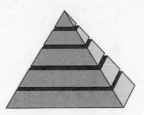

|  Figure 1  |  Figure 2  |  Figure 3  |
| :-: | :-: | :-: |
| **The Parthenon in Athens** | **The Great Pyramid at Gizeh** | **The human body** |

The earliest known example of the golden ratio in architecture is the Great Pyramid at Gizeh (Figure 2). The ratio of a side of the pyramid's square base (775.75 feet) to its original height (481.2 feet) is

$$\frac{775.75}{481.2} = 1.612\ldots.$$

Additionally, there are numerous examples of the golden ratio in nature. The French architect Le Corbusier determined that, for a human body, the distance from the head to the navel $d_1$, and the distance from the navel to the feet $d_2$, satisfy the golden ratio (Figure 3).

**MATH**   In *The Elements*, Euclid outlined a procedure for drawing a *golden rectangle*, a rectangle whose side lengths are in the golden ratio.

1. Construct a square. Let *A* denote the upper right corner, let *D* denote the lower right corner, and let *M* denote the midpoint of the bottom side.

2. Place the compass point at *M* and draw a circular arc through *A*.

3. Extend the bottom side until it intersects the arc. Call this point of intersection *F*.

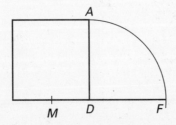

4. Draw a rectangle with base $\overline{DF}$ and height $\overline{AD}$.

5. Use a ruler to check that the sides of this rectangle satisfy the golden ratio.

NAME _____ DATE _____

# Challenge: Skills and Applications

For use with pages 488–496

1. In the diagram, $\overline{BC} \parallel \overline{DE}$.

   **a.** Write a two-column proof that $\triangle ABC \sim \triangle ADE$.

   **b.** Name another pair of similar triangles in the diagram, and write a two-column proof.

   **c.** Show that $\dfrac{AC}{AE} = \dfrac{BF}{FE}$.

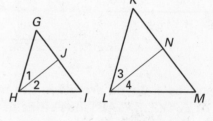

2. Write a paragraph proof.

   **Given:** $\triangle GHI \sim \triangle KLM$, $\overrightarrow{HJ}$ bisects $\angle GHI$,
   $\overrightarrow{LN}$ bisects $\angle KLM$

   **Prove:** $\dfrac{HJ}{LN} = \dfrac{GH}{KL}$

   (Hint: Prove that two triangles are similar.)

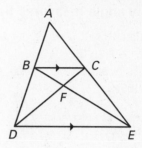

3. Write a paragraph proof.

   **Given:** $\triangle PGR \sim \triangle TUV$,
   $\overline{OG}$ is a median of $\triangle PGR$,
   $\overline{SU}$ is a median of $\triangle TUV$.

   **Prove:** $\dfrac{OG}{SU} = \dfrac{PR}{TV}$

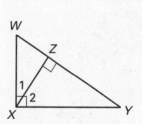

4. Write a paragraph proof.

   **Given:** $\angle WXY$ is a right angle, $\overline{XZ} \perp \overline{WY}$.

   **Prove:** $\triangle WXZ \sim \triangle XYZ$

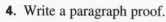

5. Refer to the diagram in Exercise 4.

   **a.** Use the result of Exercise 4 to show that $(XZ)^2 = (WZ)(ZY)$.

   **b.** Use similar triangles to show that $(WX)^2 = (WY)(WZ)$.

   **c.** Use similar triangles to show that $(XY)^2 = (WY)(ZY)$.

   **d.** Prove the Pythagorean Theorem by adding the equations from parts (c) and (d).

**LESSON**
# 8.5

## Quiz 2

**For use after Lessons 8.4 and 8.5**

**Determine whether you can show that the triangles are similar. State any angle measures that are not given.**
*(Lesson 8.4)*

**Answers**

1. _____
2. _____
3. _____
4. _____
5. _____
6. _____
7. _____

**1.**

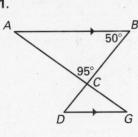

**2.**

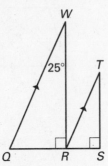

**3.**

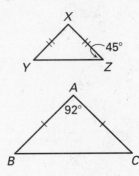

**In Exercises 4–6, you are given the ratios of the lengths of the sides of △XYZ. If △ABC has sides of lengths 4, 5, and 8 units, are the triangles similar?** *(Lesson 8.5)*

**4.** 8, 10, 16          **5.** 5, 6, 9          **6.** $6, \dfrac{15}{2}, 12$

**7. *Distance Across a Ravine*** Use the known distances in the diagram to find the distance across a ravine from *Z* to *Y*. *(Lesson 8.5)*

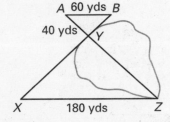

TEACHER'S NAME _____ CLASS _____ ROOM _____ DATE _____

# *Lesson Plan*

**2-day lesson** (See *Pacing the Chapter,* TE pages 454C–454D)          **For use with pages 497–505**

**GOALS**  1. **Use proportionality theorems to calculate segment lengths.**
            2. **Use proportionality theorems to solve real-life problems.**

State/Local Objectives _____

_____

## ✓ Check the items you wish to use for this lesson.

### STARTING OPTIONS
____ Homework Check: TE page 492: Answer Transparencies
____ Warm-Up or Daily Homework Quiz: TE pages 498 and 495, CRB page 83, or Transparencies

### TEACHING OPTIONS
____ Motivating the Lesson: TE page 499
____ Lesson Opener (Activity): CRB page 84 or Transparencies
____ Technology Activity with Keystrokes: CRB pages 85–86
____ Examples:   Day 1: 1–4, SE pages 498–500; Day 2: 5–6, SE page 501
____ Extra Examples:   Day 1: TE pages 499–500 or Transp.; Day 2: TE page 501 or Transp.
____ Technology Activity: SE page 497
____ Closure Question: TE page 501
____ Guided Practice: SE page 502   Day 1: Exs. 1–10; Day 2: none

### APPLY/HOMEWORK
#### Homework Assignment
____ Basic   Day 1: 11–28; Day 2: 29–39, 41–53
____ Average   Day 1: 11–28; Day 2: 29–39, 41–53
____ Advanced   Day 1: 11–28; Day 2: 29–39, 40–53

#### Reteaching the Lesson
____ Practice Masters: CRB pages 87–89 (Level A, Level B, Level C)
____ Reteaching with Practice: CRB pages 90–91 or Practice Workbook with Examples
____ Personal Student Tutor

#### Extending the Lesson
____ Applications (Interdisciplinary): CRB page 93
____ Challenge: SE page 505; CRB page 94 or Internet

### ASSESSMENT OPTIONS
____ Checkpoint Exercises:   Day 1: TE page 500 or Transp.; Day 2: TE page 501 or Transp.
____ Daily Homework Quiz (8.6): TE page 505, CRB page 97, or Transparencies
____ Standardized Test Practice: SE page 505; TE page 505; STP Workbook; Transparencies

Notes _____

_____

_____

TEACHER'S NAME _____ CLASS _____ ROOM _____ DATE _____

# Lesson Plan for Block Scheduling

1-day lesson (See *Pacing the Chapter,* TE pages 454C–454D)          For use with pages 497–505

**GOALS**   1. Use proportionality theorems to calculate segment lengths.
2. Use proportionality theorems to solve real-life problems.

State/Local Objectives _____

_____

_____

| CHAPTER PACING GUIDE | |
|---|---|
| **Day** | **Lesson** |
| 1 | Assess Ch. 7: 8.1 (all) |
| 2 | 8.2 (all); 8.3 (begin) |
| 3 | 8.3 (end); 8.4 (begin) |
| 4 | 8.4 (end); 8.5 (begin) |
| 5 | 8.5 (end); **8.6 (begin)** |
| 6 | **8.6 (end)**; 8.7 (begin) |
| 7 | 8.7 (end); Review Ch. 8 |
| 8 | Assess Ch. 8; 9.1 (all) |

✓ **Check the items you wish to use for this lesson.**

**STARTING OPTIONS**

____ Homework Check: TE page 492: Answer Transparencies

____ Warm-Up or Daily Homework Quiz: TE pages 498 and
    495, CRB page 83, or Transparencies

**TEACHING OPTIONS**

____ Motivating the Lesson: TE page 499

____ Lesson Opener (Activity): CRB page 84 or Transparencies

____ Technology Activity with Keystrokes: CRB pages 85–86

____ Examples:   Day 5: 1–4, SE pages 498–500; Day 6: 5–6, SE page 501

____ Extra Examples:   Day 5: TE pages 499–500 or Transp.; Day 6: TE page 501 or Transp.

____ Technology Activity: SE page 497

____ Closure Question: TE page 501

____ Guided Practice: SE page 502   Day 5: Exs. 1–10; Day 6: none

**APPLY/HOMEWORK**

**Homework Assignment  (See also the assignments for Lessons 8.5 and 8.7.)**

____ Block Schedule:  Day 5: 11–28; Day 6: 29–39, 41–53

**Reteaching the Lesson**

____ Practice Masters: CRB pages 87–89 (Level A, Level B, Level C)

____ Reteaching with Practice: CRB pages 90–91 or Practice Workbook with Examples

____ Personal Student Tutor

**Extending the Lesson**

____ Applications (Interdisciplinary): CRB page 93

____ Challenge: SE page 505; CRB page 94 or Internet

**ASSESSMENT OPTIONS**

____ Checkpoint Exercises:   Day 5: TE page 500 or Transp.; Day 6: TE page 501 or Transp.

____ Daily Homework Quiz (8.6): TE page 505, CRB page 97, or Transparencies

____ Standardized Test Practice: SE page 505; TE page 505; STP Workbook; Transparencies

Notes _____

_____

_____

# WARM-UP EXERCISES

For use before Lesson 8.6, pages 497–505

## Solve each proportion.

1. $\dfrac{4}{x} = \dfrac{21}{7}$

2. $\dfrac{4}{11} = \dfrac{18}{z}$

3. $\dfrac{x+1}{4} = \dfrac{6}{10}$

4. $\dfrac{3-y}{y} = \dfrac{1}{5}$

# DAILY HOMEWORK QUIZ

For use after Lesson 8.5, pages 488–496

**Are the triangles similar? If so, state the similarity and the postulate or theorem that justifies it.**

1.

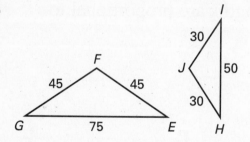

2.

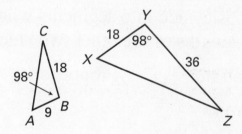

3. Find the distance across the lake shown if $CD = 120$ ft, $DX = 90$ ft, $AX = 180$ ft, and $\overline{AB} \parallel \overline{CD}$.

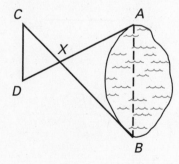

NAME _____ DATE _____

## *Activity Lesson Opener*

For use with pages 498–505

**SET UP:** Work with a partner or in a group of three.

**Three theorems, along with their diagrams and proportions, got all mixed up! Match each theorem with a diagram and a proportion to solve the puzzle in Exercise 4.**

### Theorems

1.  If a line parallel to one side of a triangle intersects the other two sides, then it divides the two sides proportionally.

    Diagram: _____ Proportion: _____

2.  If three parallel lines intersect two transversals, then they divide the transversals proportionally.

    Diagram: _____ Proportion: _____

3.  If a ray bisects an angle of a triangle, then it divides the opposite side into segments whose lengths are proportional to the lengths of the other two sides.

    Diagram: _____ Proportion: _____

### Diagrams

**A.**
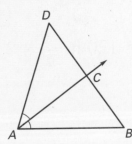

**B.**
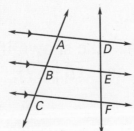

**C.**
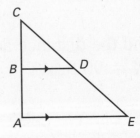

### Proportions

**D.** $\dfrac{BC}{CD} = \dfrac{AB}{AD}$   **E.** $\dfrac{AB}{BC} = \dfrac{DE}{CD}$   **F.** $\dfrac{AB}{BC} = \dfrac{DE}{EF}$

4.  Write your answers in order in the blank spaces below.

    __ __LE__RATE!  HAVE __UN!  YOU __RE __ONE!

**Geometry**
Chapter 8 Resource Book

# Technology Activity Keystrokes

**For use with page 497**

## TI-92

### Construct

**1.** Draw △*ABC*.

[F3] 3 (Place cursor at desired location for point *A*.) [ENTER] *A* (Move cursor to location for point *B*.) [ENTER] *B* (Move cursor to location for point *C*.) [ENTER] *C*

**2.** Draw point *D* on $\overline{AB}$.

[F2] 2 (Place cursor on $\overline{AB}$.) [ENTER] *D*

**3.** Draw a line through *D* that is parallel to $\overline{AC}$.

[F4] 2 (Place cursor on *D*) [ENTER] (Move cursor to $\overline{AC}$.) [ENTER]

Label the intersection of the parallel line and $\overline{BC}$ as point *E*.

[F2] 3 (Place cursor on intersection point.) [ENTER] *E*

### Investigate

**1.** Measure $\overline{BD}$, $\overline{DA}$, $\overline{BE}$, and $\overline{EC}$.

[F6] 1 (Place cursor on *B*.) [ENTER] (Move cursor to *D*.) [ENTER]

Repeat this process for the other segments.

Calculate the ratios $\dfrac{BD}{DA}$ and $\dfrac{BE}{EC}$.

[F6] 6 (Use cursor to highlight the length of $\overline{BD}$.) [ENTER] [÷] (Use cursor to highlight the length of $\overline{DA}$.) [ENTER] [ENTER] (Use cursor to highlight the length of $\overline{BE}$.) [ENTER] [÷] (Use cursor to highlight the length of $\overline{EC}$.) [ENTER] [ENTER]

**2.** Drag *DE* to different locations.

[F1] 1 (Place cursor on $\overline{DE}$.) [ENTER]

(Use the drag key 🖐 and the cursor pad to drag $\overline{DE}$.)

### Construct

**4.** Draw △*PQR*.

[F3] 3 (Place cursor at location for point *P*.) [ENTER] *P* (Move cursor to location for point *Q*.) [ENTER] *Q* (Move cursor to location for point *R*.) [ENTER] *R*

**5.** Construct the angle bisector of ∠*QPR*.

[F4] 5 (Place cursor on point *Q*.) [ENTER] (Move cursor to point *P*.) [ENTER] (Move cursor to point *R*.) [ENTER]

Label the intersection of the angle bisector and $\overline{QR}$ as point *B*.

[F2] 3 (Place cursor on the intersection point.) [ENTER] *B*

## Investigate

**5.** Measure $\overline{BR}$, $\overline{RP}$, $\overline{BQ}$, and $\overline{QP}$.

[F6] 1 (Place cursor on point *B*.) [ENTER] (Move cursor to point *R*.) [ENTER]

Repeat this process for the other segments.

Calculate the ratios $\dfrac{BR}{RP}$ and $\dfrac{RP}{QP}$.

[F6] 6 (Use cursor to highlight the length of $\overline{BD}$.) [ENTER] [÷] (Use cursor to

highlight the length of $\overline{DA}$.) [ENTER] [ENTER] (Use cursor to highlight the length

of $\overline{BE}$.) [ENTER] [÷] (Use cursor to highlight the length of $\overline{EC}$.) [ENTER] [ENTER]

## SKETCHPAD

### Construct

**1.** Draw $\triangle ABC$. Select segment from the straightedge tools and draw three segments to make up the triangle.

**2.** Draw point *D* on $\overline{AB}$ using the point tool.

**3.** Draw a line through *D* that is parallel to $\overline{AC}$. Using the selection arrow tool, select *D*, hold down the shift key, and select $\overline{AC}$. Choose **Parallel Line** from the **Construct** menu. Plot intersection point *E* of the parallel line and $\overline{BC}$ using the point tool.

### Investigate

**1.** Measure $\overline{BD}$, $\overline{DA}$, $\overline{BE}$, and $\overline{EC}$. Using the selection arrow tool, select the endpoints of a segment and then choose **Distance** from the **Measure** menu.

Calculate the ratios $\dfrac{BD}{DA}$ and $\dfrac{BE}{EC}$. Choose **Calculate** from the **Measure** menu.

Click the measure of $\overline{BD}$, click the division sign, click the measure $\overline{DA}$, and click OK. Repeat for the other ratio.

**2.** Drag $\overline{DE}$ to different locations using the translate selection arrow tool.

### Construct

**3.** Draw $\triangle PQR$. Choose segment from the straightedge tools. To relabel the points, select the text tool and double click each point.

**4.** Construct the angle bisector of $\angle QPR$. Using the selection arrow tool, select points *P*, *Q*, and *R*. Choose **Angle Bisector** from the **Construct** menu. Label the intersection of the angle bisector and $\overline{QR}$ as point *B* using the point tool.

### Investigate

**5.** Measure $\overline{BR}$, $\overline{RP}$, $\overline{BQ}$, and $\overline{QP}$. Using the selection arrow tool, select the endpoints of a segment. Choose **Distance** from the **Measure** menu.

Calculate the ratios $\dfrac{BR}{RP}$ and $\dfrac{RP}{QP}$. Choose **Calculate** from the **Measure** menu.

Click the measure of $\overline{BR}$, click the division sign, click the measure of $\overline{RP}$, and click OK. Repeat for the other ratio.

NAME _____ DATE _____

## Practice A

For use with pages 498–505

**Use the figure to complete the proportions.**

1. $\dfrac{AB}{AF} = \dfrac{BC}{?}$

2. $\dfrac{BD}{DF} = \dfrac{?}{EG}$

3. $\dfrac{AD}{BD} = \dfrac{AE}{?}$

4. $\dfrac{AC}{AG} = \dfrac{AB}{?}$

5. $\dfrac{DE}{FG} = \dfrac{AD}{?}$

6. $\dfrac{AB}{DF} = \dfrac{?}{EG}$

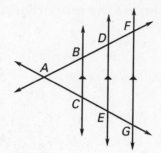

**Determine whether the statement is *true* or *false*. Explain your reasoning.**

7. $\dfrac{AB}{BD} = \dfrac{AC}{CE}$

8. $\dfrac{AC}{CE} = \dfrac{BC}{DE}$

9. $\dfrac{EC}{CA} = \dfrac{ED}{CB}$

10. $\dfrac{DB}{BA} = \dfrac{EC}{CA}$

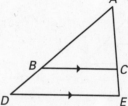

**Determine whether the given information implies $\overline{YZ} \parallel \overline{VW}$. If they are parallel, state the reason.**

11. $\dfrac{XY}{XV} = \dfrac{XZ}{XW}$

12. $\dfrac{XY}{YV} = \dfrac{XZ}{ZW}$

13. $\triangle XYZ \sim \triangle XVW$

14. $\angle VYZ \cong \angle WZY$

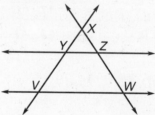

**Use the figure to match the segment with its length.**

A. 9

B. $12\frac{1}{2}$

C. 6

D. $17\frac{1}{2}$

15. $\overline{GF}$

16. $\overline{FC}$

17. $\overline{ED}$

18. $\overline{FE}$

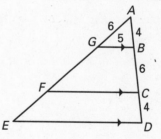

**Find the value of the variable.**

19.

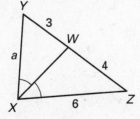

20.

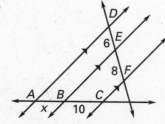

21.

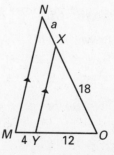

NAME _____ DATE _____

# Practice B

For use with pages 498–505

**Use the figure to complete the proportions.**

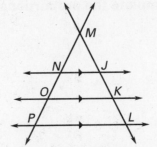

1. $\dfrac{MN}{NO} = \dfrac{MJ}{?}$

2. $\dfrac{JK}{KL} = \dfrac{?}{OP}$

3. $\dfrac{NJ}{OK} = \dfrac{MJ}{?}$

4. $\dfrac{PL}{NJ} = \dfrac{?}{MN}$

5. $\dfrac{OK}{PL} = \dfrac{MO}{?}$

6. $\dfrac{MJ}{ML} = \dfrac{?}{LP}$

**Determine whether the given information implies $\overline{BC} \parallel \overline{DE}$. Explain.**

7.

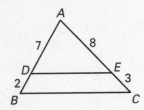

8.

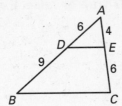

9.

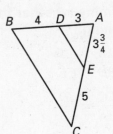

**Determine the length of each segment.**

10. $\overline{AG}$

11. $\overline{FC}$

12. $\overline{ED}$

13. $\overline{AE}$

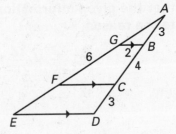

**Find the value of the variable.**

14.

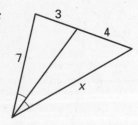

15.

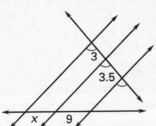

16.

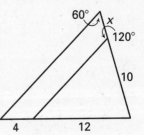

**Write a two-column or a paragraph proof.**

17. **Given:** $\overline{GB} \parallel \overline{FC} \parallel \overline{ED}$

  **Prove:** $\triangle ABG \sim \triangle ADE$

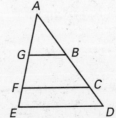

NAME _____ DATE _____

## Practice C

For use with pages 498–505

Lesson 8.6

### Use the figure to complete the proportions.

1. $\dfrac{EF}{FG} = \dfrac{BA}{?}$

2. $\dfrac{CB}{BA} = \dfrac{?}{EF}$

3. $\dfrac{EB}{FA} = \dfrac{?}{FG}$

4. $\dfrac{EG}{ED} = \dfrac{?}{CB}$

5. $\dfrac{DC}{FA} = \dfrac{?}{AG}$

6. $\dfrac{GF}{FA} = \dfrac{GD}{?}$

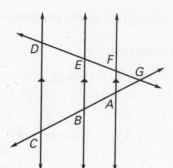

### Determine a value of the variable so that $\overline{DE} \parallel \overline{BC}$.

7.
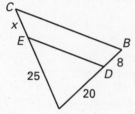

8.
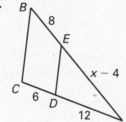

9.
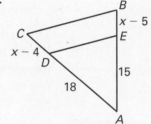

### Determine the length of each segment.

10. $\overline{AG}$

11. $\overline{FC}$

12. $\overline{ED}$

13. $\overline{AE}$

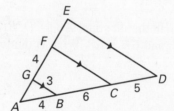

### Find the value of the variable.

14.
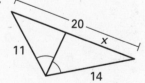

15.
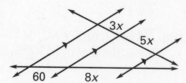

16.
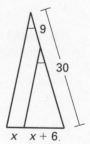

### Write a two-column or a paragraph proof.

17. **Given:** $\overline{WZ}$ bisects $\angle XZY$.
   $XW = WY$

   **Prove:** $XZ = ZY$

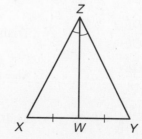

NAME _____ DATE _____

# Reteaching with Practice

For use with pages 498–505

**GOAL** **Use proportionality theorems to calculate segment lengths**

### VOCABULARY

**Theorem 8.4  Triangle Proportionality Theorem**
If a line parallel to one side of a triangle intersects the other two sides, then it divides the two sides proportionally.

**Theorem 8.5  Converse of the Triangle Proportionality Theorem**
If a line divides two sides of a triangle proportionally, then it is parallel to the third side.

**Theorem 8.6**
If three parallel lines intersect two transversals, then they divide the transversals proportionally.

**Theorem 8.7**
If a ray bisects an angle of a triangle, then it divides the opposite side into segments whose lengths are proportional to the lengths of the other two sides.

**EXAMPLE 1** *Finding the Length of a Segment*

**a.** In the diagram, $\overline{AB} \parallel \overline{CD}$, $BD = 15$, $AC = 10$, and $CE = 18$. What is the length of $\overline{DE}$?

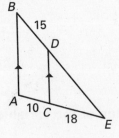

**b.** In the diagram, $\overline{NR} \parallel \overline{PQ}$, $MQ = 42$, $MN = 13$, and $NP = 8$. What is the length of $\overline{RQ}$ and $\overline{MR}$?

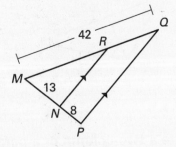

### SOLUTION

**a.** $\dfrac{BD}{AC} = \dfrac{DE}{CE}$         Triangle Proportionality Theorem

$\dfrac{15}{10} = \dfrac{DE}{18}$         Substitute.

$DE = \dfrac{15(18)}{10} = 27$         Multiply each side by 18 and simplify.

**Geometry**
Chapter 8 Resource Book

NAME _____ DATE _____

# *Reteaching with Practice*

For use with pages 498–505

**b.** Let $RQ = x$. Then $MR = 42 - x$.

$$\frac{MR}{MN} = \frac{RQ}{NP}$$     Triangle Proportionality Theorem

$$\frac{42 - x}{13} = \frac{x}{8}$$     Substitute.

$$8(42 - x) = 13x$$     Cross product property

$$336 - 8x = 13x$$     Distributive property

$$16 = x$$     Simplify.

So, $RQ = 16$ and $MR = 42 - 16 = 26$.

## *Exercises for Example 1*

### Find the value of each variable.

**1.**

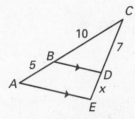

**2.**

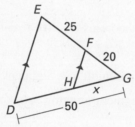

**3.**

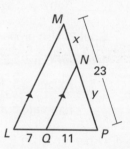

EXAMPLE 2   *Using Proportionality Theorems*

In the diagram, $\overline{MP}$ bisects $\angle M$. Find $NP$.

**SOLUTION**

$$\frac{NP}{PQ} = \frac{MN}{MQ}$$     Apply Theorem 8.7.

$$\frac{NP}{4} = \frac{16}{18}$$     Substitute.

$$NP = \frac{4(16)}{18} \approx 3.6$$     Multiply each side by 4 and simplify.

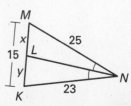

## *Exercises for Example 2*

### Find the value of each variable.

**4.**

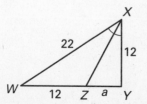

**5.**

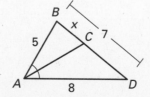

**6.**

NAME _____ DATE _____

# Quick Catch-Up for Absent Students

For use with pages 497–505

The items checked below were covered in class on (date missed) _____

### Activity 8.6: Investigating Proportional Segments (p. 497)

_____ **Goal:** Use geometry software to compare segment lengths in triangles.

_____ Student Help: Software Help

### Lesson 8.6: Proportions and Similar Triangles

_____ **Goal 1:** Use proportionality theorems to calculate segment lengths. (pp. 498–500)

*Material Covered:*

_____ Example 1: Finding the Length of a Segment

_____ Example 2: Determining Parallels

_____ Example 3: Using Proportionality Theorems

_____ Example 4: Using Proportionality Theorems

_____ Activity: Dividing a Segment into Equal Parts

_____ **Goal 2:** Use proportionality theorems to solve real-life problems. (p. 501)

*Material Covered:*

_____ Example 5: Finding the Length of a Segment

_____ Example 6: Finding Segment Lengths

_____ Other (specify) _____

_____

### Homework and Additional Learning Support

_____ Textbook (specify) pp. 502–505 _____

_____

_____ *Reteaching with Practice* worksheet (specify exercises)_____

_____ *Personal Student Tutor* for Lesson 8.6

NAME _____ DATE _____

# *Interdisciplinary Application*

For use with pages 498–505

## Harps

**MUSIC**   Although harps are known to have existed in ancient times from depictions on the Egyptian pyramids, its history and development are almost impossible to trace. As the harp spread throughout Europe and eventually to America, each culture seems to have its own version. Originally a harp could only play the notes of a scale, or the equivalent to the white piano keys. After trying to design harps with multi-rows of strings for the sharp and flat notes, a system of pedals to sharpen or flatten an entire set of notes is now a standard feature.

The harp is an instrument with almost as many shapes and sizes as there are harpists, for very little standardization exists. The types of wood, number of strings, and string spacing are literally left up to individual crafters. After almost a century of decline, a major resurgence began in the 1970s with the increased popularity of folk music. This revival continues today as harps are not only included in orchestral concerts, but also heard as background music at weddings and in restaurants and hotels.

**In Exercises 1–3, use the diagram of the first 11 strings of a 22-string folk harp shown below. Write a proportion and find the missing measure. (All lengths are in millimeters.)**

**1.** *x*

**2.** *y*

**3.** *z*

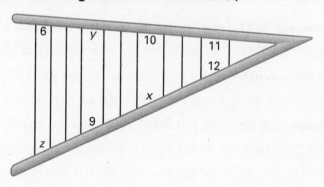

# Challenge: Skills and Applications

For use with pages 498–505

1. Figure $ABC$ is a triangle, and $D$ and $E$ are points on $\overleftrightarrow{AB}$ such that $\overrightarrow{CD}$ and $\overrightarrow{CE}$ bisect the interior and exterior angles at $C$, respectively.

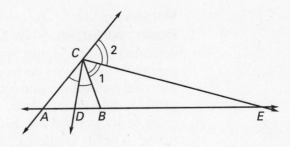

Complete the following steps to prove that $\dfrac{AD}{BD} = \dfrac{AE}{BE}$.

a. Use the fact that $\overrightarrow{CD}$ is an angle bisector to write a proportion.

b. Copy the diagram. Then draw $\overline{BF}$ such that $F$ is on $\overleftrightarrow{AC}$ and $\overleftrightarrow{BF} \parallel \overleftrightarrow{CE}$.
Use the fact that $\overleftrightarrow{BF} \parallel \overleftrightarrow{CE}$ to write a proportion involving $AE$ and $BE$.

c. Show that $\triangle FBC$ is an isosceles triangle. (*Hint:* Use the fact that $\overleftrightarrow{BF} \parallel \overleftrightarrow{CE}$.)

d. Use the results from parts (a) through (c) to show that $\dfrac{AD}{BD} = \dfrac{AE}{BE}$.

2. If $AC = BC$, is the theorem given in Exercise 1 still true? Explain.

## In Exercises 3–11, refer to the diagram in Exercise 1.

3. If $AD = 10$, $BD = 8$, and $AC = 15$, what are $BC$ and $BE$?

4. If $BE = 9$, $BD = 6$, and $BC = 8$, what are $AC$ and $AD$?

5. If $AC = 9$, $BC = 6$, and $AE = 30$, what are $AD$ and $BE$?

6. If $AB = 11$, $BC = 10$, and $BD = 5$, what are $AC$ and $BE$?

7. If $AC = 10$, $AD = 5$, and $BE = 12$, what are $BC$ and $AE$?

8. If $AD = 3x$, $AC = 4x$, $BC = x + 1$, and $BD = x$, what is $x$?

9. If $AD = x + 4$, $BC = x + 2$, $BD = x$, and $AC = BE$, what is $x$?

10. If $CD = 3$ and $CE = 8$, what is $DE$?

11. If $m\angle CAB = 40°$ and $m\angle ADC = 110°$, what are $m\angle ABC$ and $m\angle BCE$?

## *Lesson Plan*

2-day lesson (See *Pacing the Chapter*, TE pages 454C–454D)                    **For use with pages 506–514**

**GOALS**  1. **Identify dilations.**
2. **Use properties of dilations to create a real-life perspective drawing.**

State/Local Objectives _____

_____

## ✓ Check the items you wish to use for this lesson.

### STARTING OPTIONS
_____ Homework Check: TE page 502: Answer Transparencies
_____ Warm-Up or Daily Homework Quiz: TE pages 506 and 505, CRB page 97, or Transparencies

### TEACHING OPTIONS
_____ Lesson Opener (Geometry Software): CRB page 98 or Transparencies
_____ Technology Activity with Keystrokes: CRB pages 99–100
_____ Examples:   Day 1: 1–3, SE pages 506–508; Day 2: See the Extra Examples.
_____ Extra Examples:   Day 1 or Day 2: 1–3, TE pages 507–508 or Transp.
_____ Technology Activity: SE page 514
_____ Closure Question: TE page 508
_____ Guided Practice: SE page 509   Day 1: Exs. 1–7; Day 2: See Checkpoint Exs. TE pages 507–508

### APPLY/HOMEWORK
**Homework Assignment**
_____ Basic   Day 1: 8–22 even, 35, 36; Day 2: 9–23 odd, 32, 34, 38–44; Quiz 3: 1–8
_____ Average   Day 1: 8–22 even, 35, 36; Day 2: 9–23 odd, 27–32, 34, 38–44; Quiz 3: 1–8
_____ Advanced   Day 1: 8–22 even, 35, 36; Day 2: 9–23 odd, 27–32, 34, 37–44; Quiz 3: 1–8

**Reteaching the Lesson**
_____ Practice Masters: CRB pages 101–103 (Level A, Level B, Level C)
_____ Reteaching with Practice: CRB pages 104–105 or Practice Workbook with Examples
_____ Personal Student Tutor

**Extending the Lesson**
_____ Applications (Real-Life): CRB page 107
_____ Challenge: SE page 512; CRB page 108 or Internet

### ASSESSMENT OPTIONS
_____ Checkpoint Exercises:   Day 1 or Day 2: TE pages 507–508 or Transp.
_____ Daily Homework Quiz (8.7): TE page 513, or Transparencies
_____ Standardized Test Practice: SE page 512; TE page 513; STP Workbook; Transparencies
_____ Quiz (8.6–8.7): SE page 513

Notes _____

_____

_____

Lesson 8.7

TEACHER'S NAME _____ CLASS _____ ROOM _____ DATE _____

# *Lesson Plan for Block Scheduling*

1-day lesson (See *Pacing the Chapter*, TE pages 454C–454D)     **For use with pages 506–514**

**GOALS**  1. **Identify dilations.**
2. **Use properties of dilations to create a real-life perspective drawing.**

State/Local Objectives _____

_____

_____

✓ **Check the items you wish to use for this lesson.**

### STARTING OPTIONS
____ Homework Check: TE page 502: Answer Transparencies
____ Warm-Up or Daily Homework Quiz: TE pages 506 and
      505, CRB page 97, or Transparencies

### TEACHING OPTIONS
____ Lesson Opener (Geometry Software): CRB page 98 or Transparencies
____ Technology Activity with Keystrokes: CRB pages 99–100
____ Examples:   Day 6: 1–3, SE pages 506–508; Day 7: See the Extra Examples.
____ Extra Examples:   Day 6 or Day 7: 1–3, TE pages 507–508 or Transp.
____ Technology Activity: SE page 514
____ Closure Question: TE page 508
____ Guided Practice: SE page 509   Day 6: Exs. 1–7; Day 7: See Checkpoint Exs. TE pages 507–508

### APPLY/HOMEWORK
**Homework Assignment  (See also the assignment for Lesson 8.6.)**
____ Block Schedule:  Day 6: 8–22 even, 35, 36; Day 7: 9–23 odd, 27–32, 34, 38–44; Quiz 3: 1–8

**Reteaching the Lesson**
____ Practice Masters: CRB pages 101–103 (Level A, Level B, Level C)
____ Reteaching with Practice: CRB pages 104–105 or Practice Workbook with Examples
____ Personal Student Tutor

**Extending the Lesson**
____ Applications (Real-Life): CRB page 107
____ Challenge: SE page 512; CRB page 108 or Internet

### ASSESSMENT OPTIONS
____ Checkpoint Exercises:   Day 6 or Day 7: TE pages 507–508 or Transp.
____ Daily Homework Quiz (8.7): TE page 513, or Transparencies
____ Standardized Test Practice: SE page 512; TE page 513; STP Workbook; Transparencies
____ Quiz (8.6–8.7): SE page 513

| **CHAPTER PACING GUIDE** | |
|---|---|
| **Day** | **Lesson** |
| 1 | Assess Ch. 7: 8.1 (all) |
| 2 | 8.2 (all); 8.3 (begin) |
| 3 | 8.3 (end); 8.4 (begin) |
| 4 | 8.4 (end); 8.5 (begin) |
| 5 | 8.5 (end); 8.6 (begin) |
| 6 | 8.6 (end); **8.7 (begin)** |
| 7 | **8.7 (end)**; Review Ch. 8 |
| 8 | Assess Ch. 8; 9.1 (all) |

Notes _____

_____

_____

NAME _____ DATE _____

# WARM-UP EXERCISES

For use before Lesson 8.7, pages 506–514

**Find the distance between the two points.**

**1.** $A(3, 1)$, $B(5, -2)$      **2.** $G(7, -5)$, $H(8, 0)$

**Describe each type of transformation.**

**3.** a rotation

**4.** a reflection

**5.** a translation

···········································································································

# DAILY HOMEWORK QUIZ

For use after Lesson 8.6, pages 497–505

**Find the value of the variable.**

**1.**
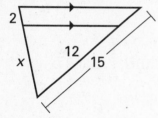

**2.**
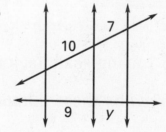

**3.** Is $\overline{GH} \parallel \overline{FE}$?

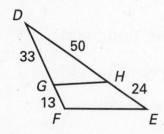

**4.** In $\triangle RST$, $\overline{RV}$ bisects $\angle TRS$. Write a proportionality statement for $\triangle RST$ based on Theorem 8.7.

NAME _____ DATE _____

## *Geometry Software Lesson Opener*

For use with pages 506–513

**Use geometry software to explore dilations. Before you begin, select the option to keep the preimage displayed.**

1. Draw a segment $\overline{AB}$ and a point $C$ not on $\overline{AB}$. Mark $C$ as the center point and dilate $\overline{AB}$ about $C$ by a scale factor of $\frac{1}{2}$. Measure lengths to find $\frac{A'C}{AC}$ and $\frac{B'C}{BC}$. How do these ratios compare with the scale factor?

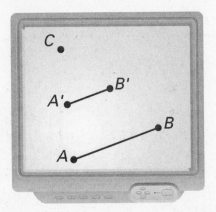

2. Repeat Exercise 1 using scale factors of $\frac{1}{3}, \frac{1}{4}, \frac{1}{5}$, and $\frac{1}{6}$. Is $\overline{AB}$ *reduced* or *enlarged* by these dilations?

3. Repeat Exercise 1 using scale factors of $\frac{3}{2}, \frac{2}{1}, \frac{5}{2}$, and $\frac{3}{1}$. Is $\overline{AB}$ *reduced* or *enlarged* by these dilations?

4. Draw a triangle $ABC$. Mark vertex $A$ as the center point and dilate $\triangle ABC$ about $A$ by a scale factor of $\frac{1}{2}$. Repeat for vertex $B$ and vertex $C$. Sketch the result and name three triangles similar to $\triangle ABC$.

5. Repeat Exercise 5 using a scale factor of $\frac{2}{1}$.

# Technology Activity Keystrokes

**For use with page 514**

## TI-92

*Construct*

1. Draw a pentagon and label it *ABCDE*.

   [F3] 4 (Place cursor at location for point *A*.) [ENTER] *A* (Move cursor to location for point *B*.) [ENTER] *B* (Move cursor to location for point *C*.) [ENTER] *C* (Move cursor to location for point *D*.) [ENTER] *D* (Move cursor to location for point *E*.) [ENTER] *E* (Move cursor back to point *A* to close pentagon.) [ENTER]

2. Draw a point outside the polygon and label it *P*.

   [F2] 1 (Place cursor at desired location for point *P*.) [ENTER] *P*

3. Set a scale factor to 0.5.

   [F7] 6 [ENTER] 0.5

   Dilate the polygon using a scale factor of 0.5 and center *P*.

   [F5] 3 (Place cursor on polygon.) [ENTER] (Move cursor to point *P*.) [ENTER] (Move cursor to scale factor.) [ENTER]

   Label the image *A'B'C'D'E'*.

   [F7] 4 (Place cursor on location of point *A'*.) [ENTER] *A* [2nd] [+] 3 7

   [F7] 4 (Place cursor on location of point *B'*.) [ENTER] *B* [2nd] [+] 3 7

   [F7] 4 (Place cursor on location of point *C'*.) [ENTER] *C* [2nd] [+] 3 7

   [F7] 4 (Place cursor on location of point *D'*.) [ENTER] *D* [2nd] [+] 3 7

   [F7] 4 (Place cursor on location of point *E'*.) [ENTER] *E* [2nd] [+] 3 7

*Investigate*

1. Measure *AP* and *A'P*.

   [F6] 1 (Place cursor on point *A*.) [ENTER] (Move cursor to point *P*.) [ENTER]

   [F6] 1 (Place cursor on point *A'*) [ENTER] (Move cursor to point *P*.) [ENTER]

   Calculate the ratio $\dfrac{AP}{A'P}$.

   [F6] 6 (Use cursor to highlight the length of *AP*.) [ENTER] [÷] (Move cursor to highlight the length of *A'P*.) [ENTER] [ENTER]

2. Measure *AB* and *A'B'*.

   [F6] 1 (Place cursor on point *A*.) [ENTER] (Move cursor to point *B*.) [ENTER]

   [F6] 1 (Place cursor on point *A'*.) [ENTER] (Move cursor to point *B'*.) [ENTER]

   Calculate the ratio $\dfrac{AB}{A'B'}$.

   [F6] 6 (Use cursor to highlight the length of *AB*.) [ENTER] [÷] (Move cursor to highlight the length of *A'B'*.) [ENTER] [ENTER]

LESSON

## 8.7

CONTINUED

NAME _____ DATE _____

# Technology Activity Keystrokes

**For use with page 514**

**3.** Drag point *P* outside polygon *ABCDE*.

    `F1` 1 (Place cursor on point *P*.) `ENTER` (Use the drag key 🖐 and the

cursor pad to drag the point.)

**4.** Drag point *P* inside polygon *ABCDE*. (See Step 3.)

**5.** Measure the areas of *ABCDE* and *A'B'C'D'E'*.

    `F6` 2 (Place cursor on polygon *ABCDE*.) `ENTER` (Place cursor on polygon

*A'B'C'D'E'*.) `ENTER`

Calculate the ratio of the area of polygon *ABCDE* to the area of polygon *A'B'C'D'E'*.

    `F6` 6 (Use cursor to highlight the area of polygon *ABCDE*.) `ENTER` `÷`

(Use cursor to highlight the area of polygon *A'B'C'D'E'*.) `ENTER` `ENTER`

## SKETCHPAD

### *Construct*

**1.** Draw pentagon *ABCDE*. Select segment from the straightedge tools.

**2.** Draw point *P* outside the polygon using the point tool.

**3.** Dilate the polygon using a scale factor of 0.5 and center *P*. Using the selection arrow tool, select *P* and choose **Mark Center** from the **Transform** menu. Select the segments and points of the polygon by holding down the shift key and selecting them. Choose **Dilate** from the **Transform** menu, enter 0.5, and click OK.

### *Investigate*

**1.** Measure *AP* and *A'P*. Using the selection arrow tool, select the endpoints of *AP*, and choose **Distance** from the **Measure** menu. Repeat this process for *A'P*. To calculate

the ratio $\dfrac{AP}{A'P}$, choose **Calculate** from the **Measure** menu. Then click the measure

of *AP*, click `÷`, click the measure of *A'P*, and click OK.

**2.** Measure *AB* and *A'B'*. Using the selection arrow tool, select the endpoints of *AB*, and choose **Distance** from the **Measure** menu. Repeat this process for *A'B'*. To calculate

the ratio $\dfrac{AB}{A'B'}$, choose **Calculate** from the **Measure** menu. Then click the measure

of *AB*, click `÷`, click the measure of *A'B'*, and click OK.

**3.** Choose the translate selection arrow tool to drag point *P* to several locations outside *ABCDE*.

**4.** Choose the translate selection arrow tool to drag point *P* to several locations inside *ABCDE*.

**5.** Measure the areas of polygons *ABCDE* and *A'B'C'D'E'*. Use the selection arrow tool to select the segments of *ABCDE*. Choose **Polygon Interior** from the **Construct** menu. Repeat this process for *A'B'C'D'E'*. Select the two polygon interiors and choose **Area** from the **Measure** menu. Calculate the ratio of the areas. Choose **Calculate** from the **Measure** menu. Click the area of *ABCDE*, click the `÷`, click the area of *A'B'C'D'E'*, and click OK.

## Practice A

**For use with pages 506–513**

**△ABC is mapped onto △A'B'C' by a dilation at D. Complete the statement.**

1. △ABC is (congruent, similar) to △A'B'C'.

2. If $\dfrac{DA}{DA'} = \dfrac{3}{5}$, then △A'B'C' is (larger, smaller) than △ABC, and the dilation is (a reduction, an enlargement).

3. If $\dfrac{DB}{DB'} = \dfrac{3}{2}$, then △A'B'C' is (larger, smaller) than △ABC, and the dilation is (a reduction, an enlargement).

**Identify the dilation and find its scale factor.**

4.
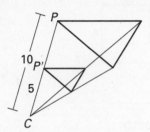

5.
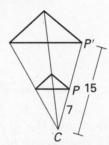

6.
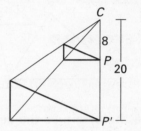

**The larger polygon is an enlargement of the smaller polygon. What is the scale factor? Solve for the variables.**

7.
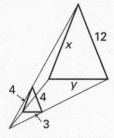

8.
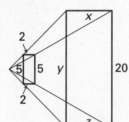

9.
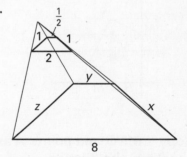

**Use the origin as the center of the dilation and the given scale factor to find the coordinates of the vertices of the image of the polygon.**

10. $k = 2$
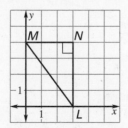

11. $k = \dfrac{1}{2}$
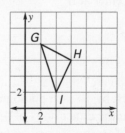

**Lesson 8.7**

NAME _____  DATE _____

# *Practice B*

For use with pages 506–513

**Identify the dilation and find its scale factor.**

**1.**

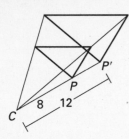

**2.**

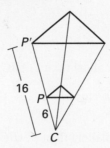

**3.**

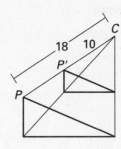

**Identify the dilation, and find its scale factor. Then, find the values of the variables.**

**4.**

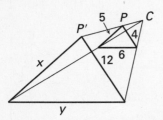

**5.**

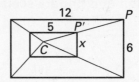

**Use the origin as the center of the dilation and the given scale factor to find the coordinates of the vertices of the image of the polygon.**

**6.** $k = 3$

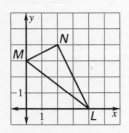

**7.** $k = \frac{1}{3}$

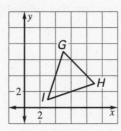

**Copy △*MNO* and points *A* and *B* as shown. Then, use a straightedge and a compass to construct the dilation.**

**8.** $k = 2$; Center at $A$

**9.** $k = \frac{1}{2}$; Center at $B$

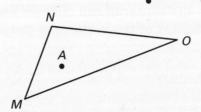

**10.** An 8-inch by 10-inch photograph is being reduced by a scale factor of $\frac{3}{4}$. What are the dimensions of the new photograph?

NAME _____ DATE _____

# *Practice C*

For use with pages 506–513

**Identify the dilation, and find its scale factor. Then, find the values of the variables.**

**1.**
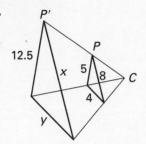

**2.**
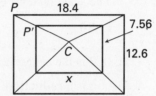

**3.**
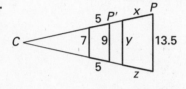

**4.**
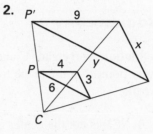

**Use the origin as the center of the dilation and the given scale factor to find the coordinates of the vertices of the image of the polygon.**

**5.** $k = \frac{2}{3}$

**6.** $k = \frac{5}{2}$

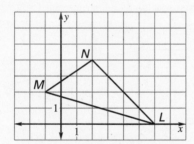

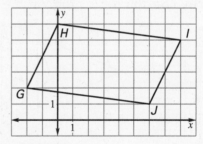

**7.** You are making hand shadows on a wall using a flashlight. You hold your hand 1 foot from the flashlight and 5 feet from the wall. Your hand is parallel to the wall. If the measure from your thumb to ring finger is 6 inches, what will be the distance between them in the shadow?

# Reteaching with Practice

**For use with pages 506–513**

**GOAL** **Identify dilations and use properties of dilations to create a perspective drawing**

> ## VOCABULARY
>
> A **dilation** with center $C$ and scale factor $k$ is a transformation that maps every point $P$ in the plane to a point $P'$ so that the following properties are true.
>
> 1. If $P$ is not the center point $C$, then the image point $P'$ lies on $\overrightarrow{CP}$.
>
>    The scale factor $k$ is a positive number such that $k = \dfrac{CP'}{CP}$, and
>
>    $k \neq 1$.
>
> 2. If $P$ is the center point $C$, then $P = P'$.
>
> A dilation is a **reduction** if $0 < k < 1$.
>
> A dilation is an **enlargement** if $k > 1$.

**EXAMPLE 1** *Identifying Dilations*

Identify the dilation and find its scale factor.

a.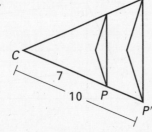

b.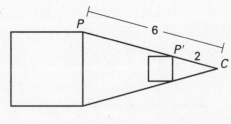

### SOLUTION

**a.** Because $\dfrac{CP'}{CP} = \dfrac{10}{7}$, the scale factor is $k = \dfrac{10}{7}$. This is an enlargement.

**b.** Because $\dfrac{CP'}{CP} = \dfrac{2}{6} = \dfrac{1}{3}$, the scale factor is $k = \dfrac{1}{3}$. This is a reduction.

### Exercises for Example 1

**Identify the dilation and find its scale factor.**

**1.**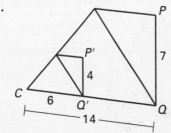

**2.**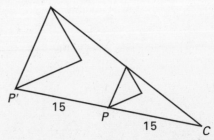

# Reteaching with Practice

For use with pages 506–513

**3.**

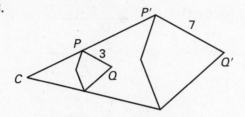

**4.**

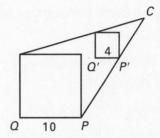

---

**EXAMPLE 2** *Dilation in a Coordinate Plane*

Draw a dilation of $\triangle ABC$ with $A(1, 2)$, $B(5, 0)$, and $C(3, 4)$. Use the origin as the center and use a scale factor of $k = 2$.

**SOLUTION**

Because the origin is the center, you can find the image of each vertex by multiplying its coordinates by the scale factor.

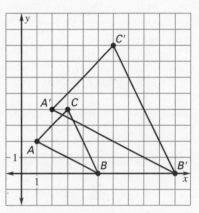

$A(1, 2) \rightarrow A'(2, 4)$

$B(5, 0) \rightarrow B'(10, 0)$

$C(3, 4) \rightarrow C'(6, 8)$

## Exercises for Example 2

**Use the origin as the center of the dilation and the given scale factor to find the coordinates of the vertices of the image of the polygon.**

**5.** $k = \dfrac{3}{2}$

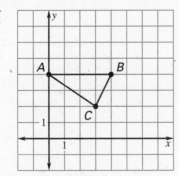

**6.** $k = 3$

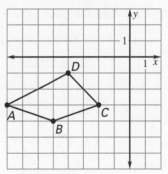

**7.** $k = \dfrac{1}{2}$

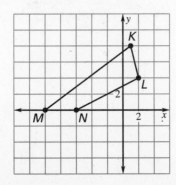

**8.** $k = \dfrac{3}{4}$

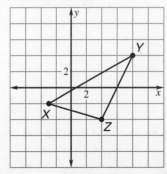

NAME _____ DATE _____

# Quick Catch-Up for Absent Students

For use with pages 506–514

The items checked below were covered in class on (date missed) _____

**Lesson 8.7: Dilations**

____ **Goal 1:** Identify dilations. (pp. 506–507)

*Material Covered:*

____ Student Help: Look Back

____ Example 1: Identifying Dilations

____ Example 2: Dilation in a Coordinate Plane

____ Activity: Drawing a Dilation

*Vocabulary:*

dilation, p. 506                                    reduction, p. 506

enlargement, p. 506

____ **Goal 2:** Use properties of dilations to create a real-life perspective drawing. (p. 508)

*Material Covered:*

____ Example 3: Finding the Scale Factor

**Activity 8.7: Exploring Dilations (p. 514)**

____ **Goal:** Use geometry software to explore properties of dilations.

____ Student Help: Software Help

____ Other (specify) _____

_____

**Homework and Additional Learning Support**

____ Textbook (specify)  pp. 509–513 _____

_____

____ *Reteaching with Practice* worksheet (specify exercises)_____

____ *Personal Student Tutor* for Lesson 8.7

NAME _____  DATE _____

# Real-Life Application: When Will I Ever Use This?

**For use with pages 506–513**

## Matreshkas

Matreshkas are Russian nesting dolls. A small doll, perhaps the size of a thimble, is placed inside a slightly larger doll. That doll is placed inside a larger doll and so on until several dolls are nested inside. The dolls are usually made from wood and twist apart around the waist to reveal the next doll. According to Russians, the wife of a Russian patron brought the idea from Japan. Some Japanese indicate that a Russian monk introduced the doll to Japan. Whatever the case, in the 1890s, a professional artist in Russia made the first sketches of Matreshka. Skilled craftsmen carve the dolls from wood. They rely on their experience rather than measurement.

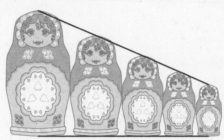

The height of the first doll is 5 inches. The height of the second doll is 4 inches. The height of each consecutive doll is scaled down with the same ratio.

  1. What is the scale factor?

  2. What is the height of the third doll?

  3. How many dolls are nested inside the first doll if the last doll is about 1 inch high?

  4. How many dolls would be nested inside the first doll if the scale factor were reduced to 75%?

# *Challenge: Skills and Applications*

To find the coordinates of a point on the image of a dilation *not* centered at the origin, you can follow the steps below.

**Step 1** Subtract the horizontal coordinate of the center from the horizontal coordinate of the point on the preimage. Then subtract the vertical coordinate of the center from the vertical coordinate of the point on the preimage.

**Step 2** Multiply the differences you found in Step 1 by the scale factor.

**Step 3** Add the horizontal and vertical coordinates of the center to the horizontal and vertical components you found in Step 2.

1. Refer to the diagram at the right. Use the steps above to draw a dilation of square *ABCD* using the center (4, 3) and a scale factor of 2.

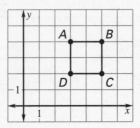

2. Let (*x*, *y*) be the coordinates of a point on the preimage, let (*a*, *b*) be the coordinates of the center of the dilation, and let *k* be the scale factor.

   **a.** Use the steps listed above and the variables *x*, *y*, *a*, *b*, and *k* to write variable expressions for the horizontal and vertical coordinates of the point on the image that corresponds to the point (*x*, *y*) on the preimage.

   **b.** Use the variable expressions you wrote in part (a) to find the coordinates of the vertices of a dilation of square *ABCD* shown above using the center (4, 3) and a scale factor of 2. Use these coordinates to draw the dilation. Does your drawing match your drawing from Exercise 1?

**In Exercises 3–8, refer to the diagram below. First find the vertices of the image after the dilation described. Then use the vertices to draw the image and preimage in the same coordinate plane.**

3. Dilate △*ABC* using center (6, 1) and scale factor 4.

4. Dilate △*ABC* using center (4, 4) and scale factor 3.

5. Dilate △*ABC* using center (−2, 13) and scale factor 2.

6. Dilate trapezoid *DEFG* using center (0, 1) and scale factor 2.

7. Dilate trapezoid *DEFG* using center (2, 3) and scale factor $\frac{3}{2}$.

8. Dilate trapezoid *DEFG* using center (0, 9) and scale factor $\frac{1}{2}$.

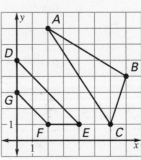

The answers to the following questions are all positive integers. List your answers and guess the next number in the sequence.

1. Consider a dilation in which both the image and the preimage are squares, the scale factor is $\frac{1}{4}$, and the area of the preimage is 1. What is the perimeter of the image?

2. Consider the following triangle where $\overline{TU}$ is parallel to $\overline{QS}$.

   What is the length of $\overline{SU}$?

   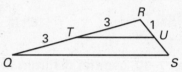

3. What is the scale factor if a square of area 4 is enlarged to a square of area 16?

4. Consider the figure at the right.

   What is the value of $x$?

   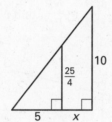

5. The two triangles at the right are similar.

   What is the length of $\overline{PR}$?

   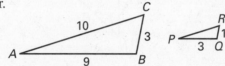

6. What is the scale factor if an isosceles right triangle of area $\frac{1}{2}$ is enlarged to an isosceles right triangle of area 32?

7. Consider the triangle below and suppose $\overline{CD}$ bisects $\angle ACB$. What is the length of $\overline{BD}$?

   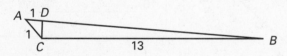

8. What is the geometric mean of 3 and 147?

9. Solve $\dfrac{17}{5} = \dfrac{x}{10}$.

10. The triangles at the right are similar.

    What is the measure of $\angle NMP$?

    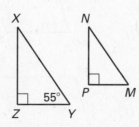

11. If the width of a golden rectangle is $\dfrac{178}{1 + \sqrt{5}}$, what is its length?

    $\left(\text{Hint: The golden ratio is } 1: \dfrac{1 + \sqrt{5}}{2}.\right)$

*Review and Assess*

**Solve the proportion.**

**1.** $\dfrac{x}{4} = \dfrac{8}{2}$    **2.** $\dfrac{25}{y} = \dfrac{5}{15}$    **3.** $\dfrac{3}{7} = \dfrac{6}{z}$    **4.** $\dfrac{4}{5} = \dfrac{x}{9}$

**Complete the sentence.**

**5.** If $\dfrac{7}{2} = \dfrac{a}{b}$, then $\dfrac{7}{a} = \dfrac{?}{b}$.    **6.** If $\dfrac{9}{n} = \dfrac{4}{m}$, then $\dfrac{9+n}{n} = \dfrac{?}{m}$.

**In Exercises 7–10, use the diagram and the given information to find the unknown length.**

**7.** Given $\dfrac{AB}{BD} = \dfrac{AC}{CE}$, find $CE$.    **8.** Given $\dfrac{MO}{OQ} = \dfrac{NP}{PQ}$, find $NP$.

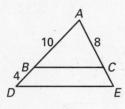

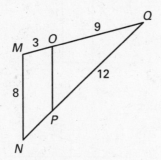

**9.** Given $\dfrac{EF}{DF} = \dfrac{FG}{FH}$, find $EF$.    **10.** Given $\dfrac{AC}{AE} = \dfrac{AB}{AD}$, find $CE$.

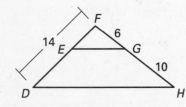

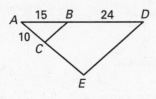

**In Exercises 11–15, *ABCD* ~ *WXYZ*.**

**11.** Find the scale factor of *ABCD* to *WXYZ*.

**12.** Find the scale factor of *WXYZ* to *ABCD*.

**13.** Find the values of *s*, *t*, and *u*.

**14.** Find the perimeter of each polygon.

**15.** Find the ratio of the perimeter of *ABCD* to the perimeter of *WXYZ*.

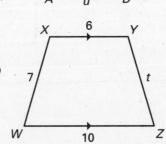

**Answers**

**1.** _____

**2.** _____

**3.** _____

**4.** _____

**5.** _____

**6.** _____

**7.** _____

**8.** _____

**9.** _____

**10.** _____

**11.** _____

**12.** _____

**13.** *s* = _____

*t* = _____

*u* = _____

**14.** *ABCD:* _____

*WXYZ:* _____

**15.** _____

*Review and Assess*

# Chapter Test A

**For use after Chapter 8**

Are the triangles similar? If so, state the similarity and a
postulate or theorem that can be used to prove that the
triangles are similar.

**16.**

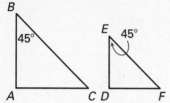

**17.**

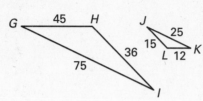

**18.**

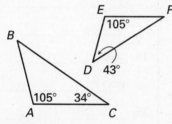

**19.**

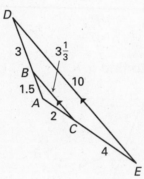

**20.**

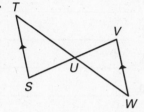

**21.**

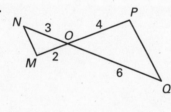

**In Exercises 22–27, use the figure shown.**

**22.** $\dfrac{YO}{OB} = \dfrac{AE}{?}$  **23.** $\dfrac{YB}{OB} = \dfrac{?}{ER}$

**24.** $\dfrac{?}{AE} = \dfrac{YB}{YO}$  **25.** $\dfrac{DY}{YO} = \dfrac{DA}{?}$

**26.** $\dfrac{DR}{?} = \dfrac{DB}{YB}$  **27.** $\dfrac{?}{AE} = \dfrac{DO}{YO}$

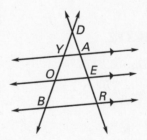

**Find the value of x.**

**28.**

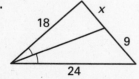

**29.**

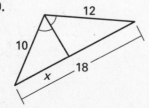

**30.**

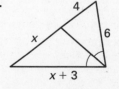

16. _____

17. _____

_____

18. _____

19. _____

20. _____

21. _____

22. _____

23. _____

24. _____

25. _____

26. _____

27. _____

28. _____

29. _____

30. _____

**Review and Assess**

NAME _____ DATE _____

# Chapter Test B

For use after Chapter 8

**Solve the proportion.**

**1.** $\dfrac{x}{3} = \dfrac{4}{2}$

**2.** $\dfrac{5}{x} = \dfrac{3}{6}$

**3.** $\dfrac{x + 2}{2} = \dfrac{3 + 3}{3}$

**4.** $\dfrac{x + 1}{2} = \dfrac{3}{4}$

**Complete the sentence.**

**5.** If $\dfrac{5}{3} = \dfrac{s}{t}$, then $\dfrac{t}{3} = $ ____?____.

**6.** If $\dfrac{2}{9} = \dfrac{m}{n}$, then $\dfrac{9}{2} = $ ____?____.

**7.** If $\dfrac{5}{p} = \dfrac{4}{q}$, then $\dfrac{5 + p}{p} = \dfrac{?}{q}$.

**In Exercises 8–10, use the diagram to find x.**

**8.**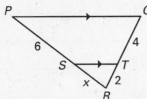

**9.**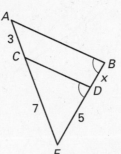

**10.**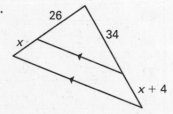

**In Exercises 11–15, ABCD ~ WXYZ.**

**11.** Find the scale factor of *ABCD* to *WXYZ*.

**12.** Find the scale factor of *WXYZ* to *ABCD*.

**13.** Find the values of *s*, *t*, and *u*.

**14.** Find the perimeter of each polygon.

**15.** Find the ratio of the perimeter of *ABCD* to the perimeter of *WXYZ*.

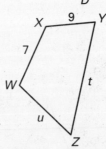

| Answers |
|---|
| **1.** _____ |
| **2.** _____ |
| **3.** _____ |
| **4.** _____ |
| **5.** _____ |
| **6.** _____ |
| **7.** _____ |
| **8.** _____ |
| **9.** _____ |
| **10.** _____ |
| **11.** _____ |
| **12.** _____ |
| **13.** *s* = _____ |
| *t* = _____ |
| *u* = _____ |
| **14.** *ABCD*: _____ |
| *WXYZ*: _____ |
| **15.** _____ |

**Are the triangles similar? If so, state the similarity and a postulate or theorem that can be used to prove that the triangles are similar.**

**16.**
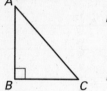

**17.**
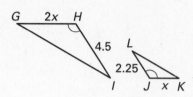

**18.**
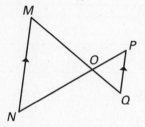

**19.**
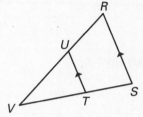

**20.**
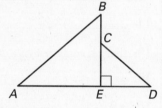

**21.**
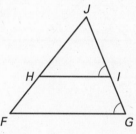

**In Exercises 22–27, use the figure shown.**

**22.** $\dfrac{AB}{CB} = \dfrac{?}{?}$   **23.** $\dfrac{AB}{BD} = \dfrac{?}{?}$

**24.** $\dfrac{GF}{GE} = \dfrac{?}{?}$   **25.** $\dfrac{CD}{AD} = \dfrac{?}{?}$

**26.** $\dfrac{DE}{GF} = \dfrac{?}{?}$   **27.** $\dfrac{GF}{ED} = \dfrac{?}{?}$

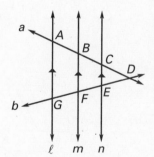

**Find the value of x.**

**28.**
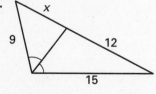

**29.**
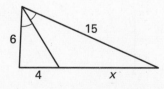

**30.**
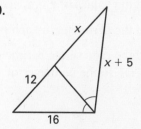

| 16. _____ |
| --- |
| _____ |
| 17. _____ |
| _____ |
| 18. _____ |
| _____ |
| 19. _____ |
| _____ |
| 20. _____ |
| _____ |
| 21. _____ |
| _____ |
| 22. _____ |
| 23. _____ |
| 24. _____ |
| 25. _____ |
| 26. _____ |
| 27. _____ |
| 28. _____ |
| 29. _____ |
| 30. _____ |

*Review and Assess*

# Chapter Test C

**For use after Chapter 8**

## Solve the proportion.

**1.** $\dfrac{3}{4} = \dfrac{x}{24}$

**2.** $\dfrac{6}{17} = \dfrac{2x}{68}$

**3.** $\dfrac{x+2}{x} = \dfrac{7}{6}$

**4.** $\dfrac{2}{5} = \dfrac{x}{x+15}$

**5.** $\dfrac{5b}{8+b} = \dfrac{4}{3}$

**6.** $\dfrac{2c-15}{7} = \dfrac{2c}{9}$

## Complete the sentence.

**7.** If $\dfrac{2}{3} = \dfrac{a}{b}$, then $\dfrac{b}{3} =$ ___?___ .

**8.** If $\dfrac{5}{8} = \dfrac{m}{n}$, then $\dfrac{10}{16} =$ ___?___ .

**9.** If $\dfrac{7}{p} = \dfrac{3}{q}$, then $\dfrac{7+p}{p} =$ ___?___ .

**10.** If $\dfrac{9+m}{m} = \dfrac{9}{3}$, then $\dfrac{9}{m} =$ ___?___ .

## In Exercises 11–13, use the diagram to find x.

**11.**

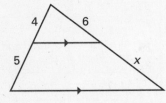

**12.**

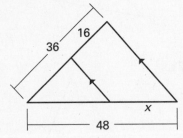

**13.**

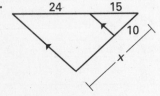

## In Exercises 14–18, ABCDE~ VWXYZ.

**14.** Find the scale factor of *ABCDE* to *VWXYZ*.

**15.** Find the scale factor of *VWXYZ* to *ABCDE*.

**16.** Find the values of *r*, *s*, *t*, and *u*.

**17.** Find the perimeter of each polygon.

**18.** Find the ratio of the perimeter of *ABCDE* to the perimeter of *VWXYZ*.

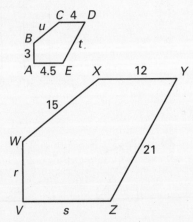

| Answers |
|---|
| 1. _____ |
| 2. _____ |
| 3. _____ |
| 4. _____ |
| 5. _____ |
| 6. _____ |
| 7. _____ |
| 8. _____ |
| 9. _____ |
| 10. _____ |
| 11. _____ |
| 12. _____ |
| 13. _____ |
| 14. _____ |
| 15. _____ |
| 16. *s =* _____ |
| *t =* _____ |
| *u =* _____ |
| 17. *ABCDE:* _____ |
| *VWXYZ:* _____ |
| 18. _____ |

*Review and Assess*

# Chapter Test C

**For use after Chapter 8**

**Are the triangles similar? If so, state the similarity and a postulate or theorem that can be used to prove that the triangles are similar.**

**19.**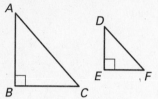

**20.**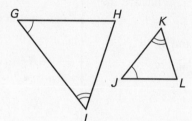

**21.**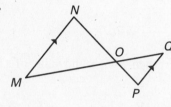

**22.**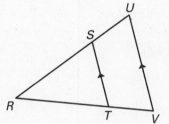

**23.**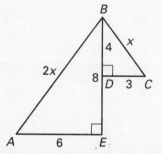

**In Exercises 24–29, use the figure shown.**

**24.** $\dfrac{AB}{CB} = \dfrac{?}{?}$  **25.** $\dfrac{AB}{BD} = \dfrac{?}{?}$

**26.** $\dfrac{GF}{GE} = \dfrac{?}{?}$  **27.** $\dfrac{CD}{AD} = \dfrac{?}{?}$

**28.** $\dfrac{DE}{GF} = \dfrac{?}{?}$  **29.** $\dfrac{GF}{ED} = \dfrac{?}{?}$

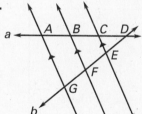

**Find the value of x.**

**30.**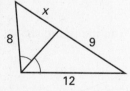

**31.**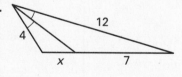

**32.**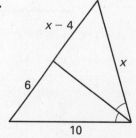

19. _____

   _____

20. _____

   _____

21. _____

   _____

22. _____

   _____

23. _____

   _____

24. _____

25. _____

26. _____

27. _____

28. _____

29. _____

30. _____

31. _____

32. _____

Review and Assess

# SAT/ACT Chapter Test

**For use after Chapter 8**

1. The perimeter of a rectangle is 36. The ratio of the lengths of the sides is 2:7. What are the lengths of the sides?

(A) 4 and 14    (B) 6 and 16

(C) 2 and 12    (D) 8 and 28

(E) 13 and 19

2. Which of the following pairs of numbers has a geometric mean of 44?

(A) 3 and 72    (B) 40 and 48

(C) 16 and 121    (D) 2 and 484

(E) 96 and 17

3. The triangles shown are similar. Which of the following is *not* a correct statement?

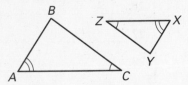

(A) $\dfrac{AB}{XY} = \dfrac{BC}{YZ}$    (B) $\triangle ABC \sim \triangle XYZ$

(C) $\dfrac{BC}{YZ} = \dfrac{AC}{XY}$    (D) $\dfrac{CA}{ZX} = \dfrac{BA}{YX}$

(E) $\dfrac{AC}{XZ} = \dfrac{AB}{XY}$

4. The two parallelograms shown are similar. What are the values of $x$ and $y$?

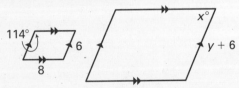

(A) $x = 114°, y = 9$   (B) $x = 66°, y = 3$

(C) $x = 114°, y = 3$   (D) $x = 66°, y = 9$

(E) $x = 114°, y = 10$

5. If $\dfrac{a}{b} = \dfrac{x}{y}$, then which one of the following is not necessarily true?

(A) $ay = bx$    (B) $\dfrac{a + x}{b + y} = \dfrac{a}{b} + \dfrac{x}{y}$

(C) $\dfrac{b}{a} = \dfrac{y}{x}$    (D) $\dfrac{y}{b} = \dfrac{x}{a}$

(E) $\dfrac{y + b}{b} = \dfrac{x + a}{a}$

6. If $\dfrac{5 + x}{x} = \dfrac{15}{6}$, then what is the value of $x$?

(A) 3    (B) $3\dfrac{1}{3}$    (C) $3\dfrac{1}{4}$

(D) $3\dfrac{1}{2}$    (E) 4

7. What is the perimeter of $\triangle ABC$?

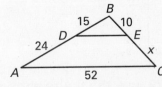

(A) 114    (B) 124    (C) 101

(D) 121    (E) 117

8. What is $CE$?

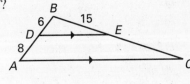

(A) 20    (B) 11.25    (C) 25

(D) 33    (E) 14

9. What is the value of $x$ in the figure shown?

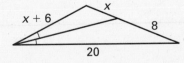

(A) 3    (B) 5    (C) 6

(D) 4    (E) 7

**JOURNAL** **1.** Use the diagram at the right showing *ABCDE ~ MIJKL*. List all the pairs of congruent angles. Write the ratios of the corresponding sides in a statement of proportionality. Find the scale factor of *MIJKL* to *ABCDE*. Find the values of *w*, *x*, *y*, and *z*. Find the perimeter of each polygon. Find the ratio of the perimeter of *ABCDE* to the perimeter of *MIJKL*.

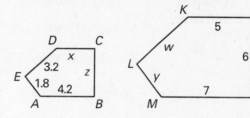

**MULTI-STEP** **2.** Use the diagram below showing △*CAB ~* △*CUP* and
**PROBLEM** △*CAB ~* △*COT*.

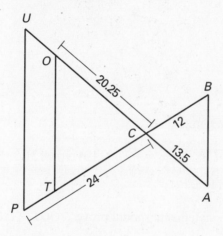

   **a.** Find the scale factor of △*CAB* to △*CUP* and △*CAB* to △*COT*.

   **b.** *CU* = __?__ .

   **c.** *CT* = __?__ .

   **d.** *OT* = __?__ .

   **e.** *UP* = __?__ .

**3.** *Critical Thinking*   Use the diagram from Exercise 2.

   **a.** Theorem 8.7 states "If a ray bisects an angle of a triangle, then it divides the opposite side into segments whose lengths are proportional to the lengths of the other two sides." Draw $\overrightarrow{AN}$ so that ∠*CAB* is bisected. Find the lengths of $\overline{CN}$ and $\overline{NB}$.

   **b.** State the Converse of the Triangle Proportionality Theorem. Use this theorem to show $\overline{UP}$ is parallel to $\overline{OT}$.

**4.** *Writing*   Use the diagram from Exercise 2. Write a paragraph proof to prove △*COT ~* △*CUP*.

*Review and Assess*

# Alternative Assessment Rubric

For use after Chapter 8

**JOURNAL**
**SOLUTION**

1. Complete answers should include:

• Congruent angles $\angle A \cong \angle M$, $\angle B \cong \angle I$, $\angle C \cong \angle J$, $\angle D \cong \angle K$, $\angle E \cong \angle L$.

• Statement of proportionality $\dfrac{AB}{MI} = \dfrac{BC}{IJ} = \dfrac{CD}{JK} = \dfrac{DE}{KL} = \dfrac{EA}{LM}$.

• Scale factor of 5 to 3.

• $w = 5.33$, $x = 3$, $y = 3$, and $z = 3.6$

• Perimeter of $ABCDE$ is 15.8 and the perimeter of $MIJKL$ is 26.33.

• The ratio of the perimeters is 3 to 5.

**MULTI-STEP**
**PROBLEM**
**SOLUTION**

2. **a.** The scale factor of $\triangle CAB$ to $\triangle CUP$ is 1 to 2 and the scale factor of $\triangle CAB$ to $\triangle COT$ is 2 to 3.

**b.** 27 units

**c.** 18 units

**d.** 22.5 units

**e.** 30 units

3. **a.** $CN = 5.7$ units, $NB = 6.3$ units

**b.** If $\dfrac{CO}{OU} = \dfrac{CT}{TP}$, then $\overline{OT} \parallel \overline{UP}$. $\dfrac{20.25}{6.75} = \dfrac{18}{6} \rightarrow 121.5 = 121.5$.
Therefore, $\overline{OT}$ is parallel to $\overline{UP}$.

4. Answers may vary, the final statements should prove $\triangle COT \sim \triangle CUP$.

**MULTI-STEP**
**PROBLEM**
**RUBRIC**

**4** Students answer all parts of the problem correctly, showing work in a step-by-step manner. Students explain the Converse of the Triangle Proportionality Theorem correctly and clearly. Students' proofs are correct and follow a logical order.

**3** Students complete the problem, but may have some minor mathematical errors. Students' explanation of the Converse of the Triangle Proportionality Theorem may have a minor error. Students' proofs are complete.

**2** Students complete the problem, but may have some mathematical errors. Students' explanation of the Converse of the Triangle Proportionality Theorem may not follow a logical order. Students' proofs do not follow a logical order.

**1** Students' answers are incomplete. Students' explanation of the Converse of the Triangle Proportionality Theorem is incomplete. Students' proof is incomplete and does not prove the final statement.

# Project: Self-Similarity

**For use with Chapter 8**

**OBJECTIVE**   **Create fractal designs.**

**MATERIALS**   plain white and colored paper, ruler, scissors, and glue

**INVESTIGATION**   In this project, you will investigate fractals. Two characteristics of fractals are *self-similarity* and *iteration*. Think of *self-similarity* as meaning that part of the whole looks like or closely resembles the whole. Think of *iteration* as a repetitive process that allows a fractal pattern to grow.

**Exploring Self-Similarity and Iteration**   Draw and analyze a fractal pattern.

1. Begin by drawing a 9-inch line segment and label it I(0).

2. For the next iteration draw a line segment below the previous segment but with the middle third of the segment missing. Label this row I(1).

3. Continue the iteration process four more times, leaving out the middle third of *each* segment, and label each row of segments appropriately.

4. Assume the beginning segment, I(0), is one unit long. Make a table up to I(5) like the one shown. Then include a row for the general pattern, I(n).

| Iteration | Length of each segment | No. of segments in row | Total length shown in row |
|-----------|------------------------|------------------------|---------------------------|
| 0 | 1 | 1 | 1 |

5. How is self-similarity evident in this fractal pattern, known as *Cantor dust*?

**Creating a Fractal Card**

Step 1   Fold a sheet of paper in half, vertically.

Step 2   Mark two points along the folded edge, each about one-fourth the distance from one end to the other. Make a cut from each point, perpendicular to the fold, that extends half way across the paper.

Step 3   Fold that middle section over and crease it. Open the paper and push that middle section to the inside. Fold the paper closed.

Step 4   Repeat the process of Steps 2 and 3 several times, each time cutting into the remaining middle portion of the folded paper.

Step 5   Once the folding and cutting is complete, open the paper, push out the cut shape from the backside of the paper and crease firmly.

Step 6   Complete your card by gluing a piece of folded paper to the back of your fractal paper. You will have created a fractal pop-up card.

Create a fractal pop-up card of your own design that demonstrates self-similarity and iteration. Write the rule for it.

**PRESENT YOUR RESULTS**   Your report should contain the sketch and the completed table of iterations for *Cantor dust,* two fractal pop-up cards, and the rule for the card with your design. Explain how your card demonstrates self-similarity and iteration.

*Review and Assess*

# Project: Teacher Notes

**For use with Chapter 8**

**GOALS** • Investigate the concept of similarity by working with fractal patterns.

• Generalize a pattern.

**MANAGING THE PROJECT** You might wish to try both parts of this project yourself before beginning it in class. Consider showing these fractal card diagrams on the board or overhead projector. The last sketch is for your benefit.

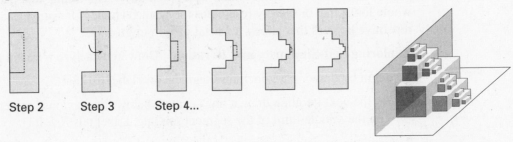

Step 2        Step 3        Step 4...

Finished card

Once the fractal pattern is cut and folded, ask students to view it from all angles and sides, and discuss it with each other. Ideas for a new pattern can come just from experimenting with cuts and folds, as well as from thinking about a particular theme, holiday, or occasion. Consider a display of the original student cards. Some fractals might be hung by string, thread or fishing line. Let students share their *iteration* processes with each other.

**RUBRIC** **The following rubric can be used to assess student work.**

**4** The student neatly and accurately completes the sketch and the table of data for *Cantor dust*, and two fractal pop-up cards. The original pop-up card is creative and unusual. The student writes an accurate iteration rule for his or her own fractal design. The student includes a detailed assessment of how his or her design demonstrates *self-similarity* and *iteration*.

**3** The student completes the sketch and the table of data for *Cantor dust*, and two fractal pop-up cards, but there may be errors in the table of data. The student writes an accurate iteration rule for his or her own fractal design. The student includes a short assessment of how his or her design demonstrates *self-similarity* and *iteration*.

**2** The student completes the sketch for *Cantor dust*, but the data in the table contains errors or is missing. The student completes two fractal pop-up cards, but the iteration rule for the original design contains errors or is missing. The student gives a very sketchy explanation of how his or her design displays *self-similarity* and *iteration*.

**1** The student completes the sketch for *Cantor dust*, but the data in the table contains errors or is missing. The student attempts the first fractal pop-up card and makes a very poor attempt to make an original design. The rule and discussion for his or her own design are missing or very sketchy.

NAME _____ DATE _____

# Cumulative Review

For use after Chapters 1–8

**Find the coordinates of the midpoint of $\overline{AB}$. (1.5)**

**1.** $A(-3, 5), B(7, 9)$

**2.** $A(0, 6), B(3, -2)$

**Name the property that justifies the statement. (2.4)**

**3.** If $AB = 8$, then $AB + 6 = 14$.

**4.** If $JK = LM$, then $LM = JK$.

**Find the value of $x$ and $y$. (3.3)**

**5.**

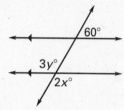

**6.**

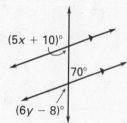

**Find the measure to the exterior angle shown. (4.1)**

**7.**

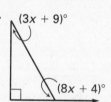

**8.**

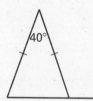

**Complete the statement given that $J$, $K$, and $L$ are midpoints of sides of $\triangle XYZ$. (5.4)**

**9.** $\overline{JK} \parallel$ ____?____

**10.** If $\overline{KL} = 6$, then $\overline{YX} =$ ____?____.

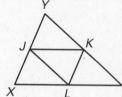

**Find the value of $x$ in each trapezoid. (6.5)**

**11.**

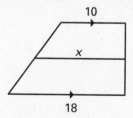

**12.**

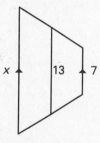

**Find the coordinates of the reflection using the preimage below. (7.2)**

**13.** reflection in the $x$-axis

**14.** reflection in the $y$-axis

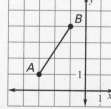

NAME_____ DATE _____

# *Cumulative Review*

For use after Chapters 1–8

**Solve the proportion. (8.1)**

15. $\dfrac{7}{10} = \dfrac{x}{15}$

16. $\dfrac{2x + 3}{4} = \dfrac{5}{6}$

**Use the diagram and the given information to find the unknown length. (8.2)**

17. $\dfrac{JN}{NK} = \dfrac{MO}{OL}$

Find *NK*.

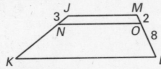

**The two polygons are similar. Find the scale factor. (8.3)**

18.

19.

**The triangles are similar. Find the values of the variable. (8.4)**

20.

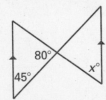

21.

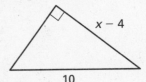

**Are the triangles similar? If so, state the similarity and the postulate or theorem that justifies your answer. (8.5)**

22.

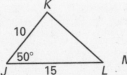

23.

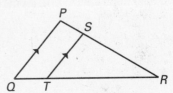

**Find the value of the variable. (8.6)**

24.

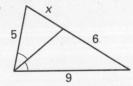

25.

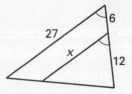

**Identify the dilation and find the scale factor. (8.7)**

26.

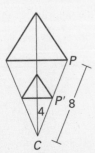

27.

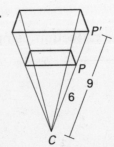

**Geometry**
Chapter 8 Resource Book

# ANSWERS

## Chapter Support

### Parent Guide

**8.1:** $\frac{5}{6}$  **8.2:** about 305 seats  **8.3:** 36 m

**8.4:** Angle-Angle Similarity Postulate  **8.5:** 90 ft

**8.6:** 50.4 m and 39.6 m

**8.7:** $A'(-4, 0)$, $B'(2, 6)$, $C'(4, -2)$

### Prerequisite Skills Review

**1.** 30 in.  **2.** 34 ft  **3.** 30 m  **4.** 36 cm

**5.** 48 yd  **6.** 54 in.  **7.** SSS Congruence
Postulate, $\triangle JKM \cong \triangle LKM$  **8.** AAS
Congruence Theorem, $\triangle CGF \cong \triangle EGD$

**9.** ASA Congruence Postulate,
$\triangle MPO \cong \triangle ONM$  **10.** HL Congruence
Theorem, $\triangle WXV \cong \triangle YZV$  **11.** ASA
Congruence Postulate, $\triangle AFD \cong \triangle CEB$

**12.** SAS Congruence Postulate, $\triangle RHU \cong \triangle RTS$

**13.** $\frac{1}{3}$  **14.** $-\frac{3}{8}$  **15.** $-\frac{2}{5}$  **16.** 0  **17.** $\frac{3}{5}$

**18** undefined

### Strategies for Reading Mathematics

**1.**   **2.** Change the label of the width
from "2 in." to "3 in."; 4.5 in.

**3.**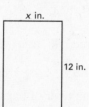

**4.** Label the length of the enlargement "$x$ in." and
label the width "12 in."; 18 in.

## Lesson 8.1

### Warm-Up Exercises

**1.** $\frac{4}{5}$  **2.** $\frac{1}{4}$  **3.** $\frac{7}{2}$  **4.** $\frac{1}{5}$

### Daily Homework Quiz

**1.** T  **2.** TV  **3.**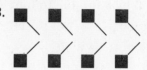

### Lesson Opener

Allow 5 minutes.

**1–12.** Answers will vary.

### Practice A

**1.** $\frac{5}{1}$  **2.** $\frac{4}{1}$  **3.** $\frac{5}{6}$  **4.** $\frac{4}{5}$  **5.** girls' team  **6.** $\frac{1}{2}$

**7.** $\frac{2}{1}$  **8.** $\frac{2}{3}$  **9.** $\frac{4}{3}$  **10.** $\frac{16 \text{ cm}}{12 \text{ cm}}; \frac{4}{3}$  **11.** $\frac{8 \text{ in.}}{10 \text{ in.}}; \frac{4}{5}$

**12.** $\frac{24 \text{ in.}}{8 \text{ in.}}; \frac{3}{1}$  **13.** $\frac{72 \text{ in.}}{24 \text{ in.}}; \frac{3}{1}$  **14.** $\frac{60 \text{ mm}}{10 \text{ mm}}; \frac{6}{1}$

**15.** $\frac{40 \text{ g}}{1000 \text{ g}}; \frac{1}{25}$  **16.** $\frac{20 \text{ ft}}{9 \text{ ft}}; \frac{20}{9}$  **17.** $\frac{48 \text{ oz}}{12 \text{ oz}}; \frac{4}{1}$

**18.** $\frac{35 \text{ days}}{30 \text{ days}}; \frac{7}{6}$  **19.** $\frac{85 \text{ cm}}{50 \text{ cm}}; \frac{17}{10}$

**20.** $\frac{10,560 \text{ ft}}{60 \text{ ft}}; \frac{176}{1}$  **21.** 2  **22.** 4  **23.** 80

**24.** $-1$  **25.** $\frac{5}{2}$  **26.** 9  **27.** $\frac{160}{1}$

### Practice B

**1.** $\frac{1}{3}$  **2.** $\frac{12}{7}$  **3.** $\frac{1}{2}$  **4.** $\frac{4}{3}$  **5.** $\frac{2 \text{ qt}}{16 \text{ qt}}; \frac{1}{8}$

**6.** $\frac{250 \text{ mg}}{10,000 \text{ mg}}; \frac{1}{40}$  **7.** $\frac{24 \text{ oz}}{32 \text{ oz}}; \frac{3}{4}$  **8.** $\frac{14 \text{ ft}}{18 \text{ ft}}; \frac{7}{9}$

**9.** $\frac{48 \text{ in.}}{8 \text{ in.}}; \frac{6}{1}$  **10.** $\frac{96 \text{ hr}}{36 \text{ hr}}; \frac{8}{3}$  **11.** $\frac{150 \text{ cm}}{80 \text{ cm}}; \frac{15}{8}$

**12.** $\frac{440 \text{ yd}}{3520 \text{ yd}}; \frac{1}{8}$  **13.** $\frac{1}{2}$  **14.** $\frac{1}{3}$  **15.** $\frac{3}{5}$  **16.** $\frac{9}{2}$

**17.** 2.25  **18.** 6  **19.** 85  **20.** 9  **21.** 25

**22.** 21  **23.** 8  **24.** 12  **25.** 27.5  **26.** 35 in.

**27.** 80 ft

### Practice C

**1.** $\frac{96 \text{ hr}}{16 \text{ hr}}; \frac{6}{1}$  **2.** $\frac{18 \text{ yd}}{2 \text{ yd}}; \frac{9}{1}$  **3.** $\frac{600 \text{ m}}{200 \text{ m}}; \frac{3}{1}$

**4.** $\frac{80 \text{ mm}}{6 \text{ mm}}; \frac{40}{3}$  **5.** $\frac{20 \text{ qt}}{4 \text{ qt}}; \frac{5}{1}$  **6.** $\frac{18 \text{ in.}}{108 \text{ in.}}; \frac{1}{6}$

# Lesson 8.1 *continued*

**7.** $\dfrac{25 \text{ min}}{120 \text{ min}}$, $\dfrac{5}{24}$  **8.** $\dfrac{220 \text{ yd}}{5280 \text{ yd}}$, $\dfrac{1}{24}$  **9.** 2

**10.** 3.2  **11.** 20  **12.** 12  **13.** 62  **14.** 6

**15.** 12 ft by 8 ft  **16.** 16 ft by 12 ft

**17.** 45°, 60°, 75°  **18.** 24°, 60°, 96°

**19.** $AB = AC = 8, BC = 4$

**20.** $AB = 6, AC = 4, BC = 8$

**21.** $\dfrac{5 \text{ in.}}{18 \text{ ft}} = \dfrac{5}{216}$  **22.** $3\frac{1}{3}$ in.

## Reteaching with Additional Practice

**1.** $\frac{1}{8}$  **2.** $\frac{2}{1}$  **3.** $\frac{1}{1}$  **4.** $\frac{2}{3}$  **5.** length = 25 ft,
width = 5 ft  **6.** 31.5  **7.** 2  **8.** $-12$  **9.** 6

## Real-Life Application

**1.** 3:4.5  **2.** Flour: 4.5 cups; Salt: 1.5 tblsp;
Yeast: 1.5 tsp; Sugar: $\frac{3}{16}$ cup; Water: 1.5 cup;
Olive oil: 6 tblsp; Tomato: 15 oz; Basil: 3 tblsp;
Oregano: 6 tblsp; Garlic powder: 6 tblsp

**3.** no  **4.** 2  **5.** $3\frac{3}{4}$ cups

## Challenge: Skills and Applications

**1.** $\dfrac{1 \text{ mi}}{5280 \text{ ft}}$, $\dfrac{5280 \text{ ft}}{1 \text{ mi}}$  **2.** $\dfrac{1 \text{ ton}}{2000 \text{ lb}}$, $\dfrac{2000 \text{ lb}}{1 \text{ ton}}$

**3.** $\dfrac{10 \text{ mm}}{1 \text{ cm}}$, $\dfrac{1 \text{ cm}}{10 \text{ mm}}$  **4.** $\dfrac{1000 \text{ m}}{1 \text{ km}}$, $\dfrac{1 \text{ km}}{1000 \text{ m}}$

**5.** $\dfrac{1 \text{ lb}}{16 \text{ oz}}$, $\dfrac{16 \text{ oz}}{1 \text{ lb}}$  **6.** $\dfrac{1 \text{ km}}{0.621 \text{ mi}}$, $\dfrac{0.621 \text{ mi}}{1 \text{ km}}$

**7.** 3700 m  **8.** 0.0675 ton  **9.** 56.7 cm

**10.** 20.25 lb  **11.** 10.5 ft  **12.** $\approx 186.3$ mi

**13.** 67,056 ft  **14.** 1,555,200 sec

**15.** $\approx 515.3$ km  **16.** $\approx 7514$ mi  **17.** 640 min

**18.** $\approx 5.49$ mi  **19.** 7,200,000 oz

# Lesson 8.2

## Warm-Up Exercises

**1.** $\frac{5}{2}$  **2.** $\frac{45}{2}$  **3.** $-\frac{3}{2}$  **4.** 3

## Daily Homework Quiz

**1.** 12 ft by 20 ft  **2.** 20°, 60°, 100°  **3.** $\frac{12}{5}$  **4.** $\frac{3}{2}$

## Lesson Opener

Allow 10 minutes.

**1.** $x = \sqrt{a \cdot b}$

**2.**

| $a$ | $b$ | $x$ |
|---|---|---|
| 5 | 20 | 10 |
| 2 | 13 | $\approx 5.1$ |
| 6 | 8 | $\approx 6.9$ |
| 12 | 3 | 6 |
| 11 | 11 | 11 |
| 1 | 9 | 3 |
| 30 | 7 | $\approx 14.5$ |
| 9 | 12 | $\approx 10.4$ |
| 2 | 8 | 4 |
| 2 | 32 | 8 |
| 1 | 25 | 5 |
| 14 | 14 | 14 |
| 45 | 5 | 15 |
| 4 | 9 | 6 |

**3.** The product $a \cdot b$ must be a perfect square.

**4.** 1 and 144, 2 and 72, 3 and 48, 6 and 24, 8 and 18, 9 and 16, 12 and 12

**5.** *Sample answer:* $x = 2$; $x$ must be a prime number.  **6.** *Sample answer:* $x = 10$; 1 and 100, 2 and 50, 4 and 25, 5 and 20, 10 and 10

## Practice A

**1.** $\dfrac{4}{3}$  **2.** $\dfrac{b}{4}$  **3.** $\dfrac{3+4}{4}$  **4.** $\dfrac{a+b}{b}$  **5.** false

**6.** false  **7.** true  **8.** true  **9.** 6  **10.** 8

**11.** 6  **12.** 10  **13.** $4\sqrt{2}$  **14.** $6\sqrt{2}$  **15.** 6

**16.** 4  **17. a.** number of bags needed
**b.** square feet of lawn  **c.** one bag  **d.** square feet per bag; 5 bags  **18. a.** your tax  **b.** value of your house  **c.** neighbor's tax  **d.** value of neighbor's house; $\approx \$1421$

## Practice B

**1.** $\dfrac{8}{5}$  **2.** $\dfrac{q}{8}$  **3.** $\dfrac{5+8}{8}$  **4.** $\dfrac{p+q}{q}$  **5.** true

**6.** false  **7.** true  **8.** false  **9.** $2\sqrt{15}$
**10.** $4\sqrt{6}$  **11.** $2\sqrt{30}$  **12.** $5\sqrt{6}$  **13.** $8\sqrt{3}$

**Geometry**
Chapter 8 Resource Book

# Lesson 8.2 *continued*

**14.** $4\sqrt{30}$ **15.** $\dfrac{24}{7}$ **16.** $\dfrac{48}{5}$

**17.** \$47.27 **18.** \$20.40 **19.** 40 feet

## Practice C

**1.** $\dfrac{9}{5}$ **2.** $\dfrac{n}{9}$ **3.** $\dfrac{5+9}{9}$ **4.** $\dfrac{m+n}{n}$ **5.** $4\sqrt{6}$

**6.** $\sqrt{105.4}$ **7.** $6\sqrt{10}$ **8.** $6\sqrt{15}$ **9.** $2a$

**10.** $2a\sqrt{2}$ **11.** $\dfrac{14}{5}$ **12.** 21.6 **13.** $-5$ **14.** 5

**15.** 800 **16.** $\approx 260$ miles

## Reteaching with Additional Practice

**1.** 3.2 **2.** 4 **3.** 9 **4.** 3

## Interdisciplinary Application

**1.** 408.17 sq in. **2.** 1355.30 sq in.

**3.** 1694 sq in. **4.** 41.2 in. $\times$ 41.2 in.

**5.** 1:753

## Challenge: Skills and Applications

**1.** 12 **2.** 4 **3.** $\dfrac{1}{2}$ **4.** 16 **5.** 10 **6.** $\dfrac{5}{3}$

**7.** $\dfrac{x(x+1)}{x+2}$ **8.** $\dfrac{2x+4}{3}$ **9.** $\dfrac{(2x+3)(x+1)}{x+3}$

**10.** 4 **11.** 4 **12.** $\dfrac{5}{2}$, 5 **13.** 50 ft

# Lesson 8.3

## Warm-Up Exercises

**1.** 122° **2.** 119° **3.** 8 **4.** $\dfrac{10}{3}$

## Daily Homework Quiz

**1.** false **2.** true **3.** 15 **4.** 12.5

## Lesson Opener

Allow 10 minutes.

**1.** 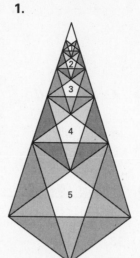 **2.**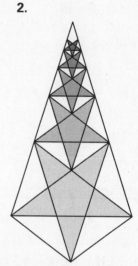

**3.** 5 pentagons, 20 obtuse isosceles triangles, 25 acute triangles, and 1 kite

## Practice A

**1.** a and c **2.** a and b **3.** a and c

**4.** a and b

**5.** $\angle G \cong \angle T$, $\angle R \cong \angle F$, $\angle M \cong \angle D$, $\dfrac{GR}{TF} = \dfrac{RM}{FD} = \dfrac{GM}{TD}$ **6.** $\angle S \cong \angle J$, $\angle T \cong \angle K$, $\angle R \cong \angle L$, $\dfrac{ST}{JK} = \dfrac{TR}{KL} = \dfrac{SR}{JL}$

**7.** $\angle A \cong \angle M$, $\angle B \cong \angle N$, $\angle C \cong \angle O$, $\angle D \cong \angle P$, $\dfrac{AB}{MN} = \dfrac{BC}{NO} = \dfrac{CD}{OP} = \dfrac{AD}{MP}$

**8.** yes; $\triangle XAR \sim \triangle MNT$ **9.** no **10.** 2:1

**11.** 1:2 **12.** $x = 16$, $y = 4$ **13.** perimeter $ABCD = 46$, perimeter $GHIJ = 23$ **14.** 2:1

## Practice B

**1.** $\angle J \cong \angle R$, $\angle K \cong \angle S$, $\angle L \cong \angle T$, $\dfrac{JK}{RS} = \dfrac{KL}{ST} = \dfrac{JL}{RT}$ **2.** $\angle T \cong \angle M$, $\angle U \cong \angle N$, $\angle V \cong \angle O$; $\dfrac{TU}{MN} = \dfrac{UV}{NO} = \dfrac{TV}{MO}$ **3.** $\angle W \cong \angle D$, $\angle X \cong \angle E$, $\angle Y \cong \angle F$, $\angle Z \cong \angle G$, $\dfrac{WX}{DE} = \dfrac{XY}{EF} = \dfrac{YZ}{FG} = \dfrac{WZ}{DG}$ **4.** no

**5.** yes; $\triangle LMN \sim \triangle TPO$ **6.** $\dfrac{5}{4}$ **7.** $\dfrac{4}{5}$ **8.** 4

# Lesson 8.3 continued

**9.** 55° **10.** 12.8 **11.** $\frac{5}{4}$ **12.** 9, 12 **13.** 13.5, 16 **14.** 28.8 in.

## Practice C

**1.** $\angle S \cong \angle C$, $\angle T \cong \angle D$, $\angle U \cong \angle E$;

$\frac{ST}{CD} = \frac{TU}{DE} = \frac{SU}{CE}$ **2.** $\angle L \cong \angle G$, $\angle M \cong \angle H$,

$\angle N \cong \angle I$; $\frac{LM}{GH} = \frac{MN}{HI} = \frac{LN}{GI}$ **3.** $\angle Q \cong \angle A$,

$\angle R \cong \angle B$, $\angle S \cong \angle C$, $\angle T \cong \angle D$;

$\frac{QR}{AB} = \frac{RS}{BC} = \frac{ST}{CD} = \frac{QT}{AD}$ **4.** $\frac{2}{3}$ **5.** $\frac{3}{2}$ **6.** 4.5

**7.** 63° **8.** 24 **9.** $\frac{3}{2}$ **10.** yes; $1:\sqrt{2}$

**11.** yes; 5:8 **12.** 69°, 12.5 **13.** 72°, 11

**14.** 60 in. **15.** 4:1; 4.5 cm, 3 cm

## Reteaching with Additional Practice

**1.** $\angle A \cong \angle D$, $\angle B \cong \angle E$, $\angle C \cong \angle F$;

$\frac{AB}{DE} = \frac{BC}{EF} = \frac{CA}{FD}$

**2.** $\angle A \cong \angle Z$, $\angle B \cong \angle W$, $\angle D \cong \angle X$, $\angle C \cong \angle Y$;

$\frac{AB}{ZW} = \frac{BD}{WX} = \frac{DC}{XY} = \frac{CA}{YZ}$

**3.** $\angle E \cong \angle M$, $\angle F \cong \angle R$, $\angle G \cong \angle Q$, $\angle H \cong \angle P$, $\angle J \cong \angle N$;

$\frac{EF}{MR} = \frac{FG}{RQ} = \frac{GH}{QP} = \frac{HJ}{PN} = \frac{JE}{NM}$

**4.** Yes, $ABCDE \sim HJKFG$ **5.** no **6.** 11.2

**7.** 14

## Real-Life Application

**1.** $\frac{1.5}{50}$ **2.** 2.9 inches high and 2.5 inches wide

**3.** 5.6 feet high and 8.3 feet wide

**4.** Answers will vary depending on the size of window the students pick.

## Challenge: Skills and Applications

**1.** $\frac{1 + \sqrt{5}}{2} \approx 1.618$ **2. a.** $WX = \frac{rv}{u}$, $XY = \frac{sv}{u}$,

and $YZ = \frac{tv}{u}$ **b.** $\frac{WX + XY + YZ + ZW}{KL + LM + MN + NK} =$

$\dfrac{\dfrac{rv}{u} + \dfrac{sv}{u} + \dfrac{tv}{u} + v}{r + s + t + u} = \dfrac{\dfrac{v}{u}(r + s + t + u)}{r + s + t + u} = \dfrac{v}{u}$.

This is the same as the ratio of any pair of corresponding sides. **3.** Since $\overline{BA} \parallel \overline{CD}$,

$\angle A \cong \angle DCE$ (Alternate Interior Angles Theorem). Since $\angle B$ and $\angle D$ are right angles, $\angle B \cong \angle D$. By the Third Angles Theorem, $\angle ACB \cong \angle CED$. So, the corresponding angles are congruent. Let $k = \frac{CD}{AB}$. Then $CD = k \cdot AB$ and $DE = k \cdot BC$. Using the Pythagorean Theorem,

$CE = \sqrt{(CD)^2 + (DE)^2}$

$= \sqrt{(k \cdot AB)^2 + (k \cdot BC)^2}$

$= \sqrt{k^2 \cdot [(AB)^2 + (BC)^2]}$

$= k \cdot \sqrt{(AB)^2 + (BC)^2} = k \cdot \sqrt{(AC)^2} = k \cdot AC.$

Therefore, $\frac{CE}{AC} = k$, so all of the corresponding side lengths are proportional. Hence, $\triangle ABC \sim \triangle CDE$. **4.** 6 **5.** $-5, \frac{5}{2}$ **6.** 1440 in.²

## Quiz 1

**1.** 18 **2.** 8 **3.** $\frac{4}{5}$ **4.** 25 **5.** $\sqrt{72} \approx 8.48$

**6.** $\sqrt{108} \approx 10.39$ **7.** 1; 2:3; $\frac{2}{3}$ **8.** 32; 1:2; $\frac{1}{2}$

**9.** yes **10.** No; the corresponding angles are all congruent. Only the corresponding lengths are 60 : 1.

# Lesson 8.4

## Warm-Up Exercises

**1.** 86° **2.** 62° **3.** 32° **4.** 20

## Daily Homework Quiz

**1.** no **2.** yes; *Sample answer: ABCD ~ FEHG*

**3.** 2:3 **4.** 12 **5.** 30°

## Lesson Opener

Allow 20 minutes.

**1.** $\triangle BCD \sim \triangle HED$; $\angle BCD \cong \angle HED$, $\angle BDC \cong \angle HDE$, $\angle CBD \cong \angle EHD$;

$\frac{BC}{HE} = \frac{CD}{ED} = \frac{BD}{HD}$

# Lesson 8.4 *continued*

**2.** *Sample answer:*

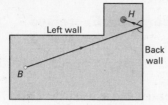

**3.** *Sample answer:*

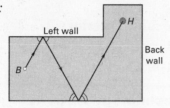

**4.** Check drawings.

## Technology Activity

**1.** The ratios are equal.  **2. a.** yes  **b.** no

**c.** no  **3. a.** $TP = 33$ and $OB = 7.67$

**b.** $CO = 7$ and $IP = 18$

## Practice A

**1.** A and B; $\triangle EFG$ is a 30°-60°-90° triangle.

**2.** A and B; $\triangle EFG$ is a 40°-70°-70° triangle.

**3.** not similar  **4.** similar  **5.** cannot be deter-

mined  **6.** $\angle Q \cong \angle T, \angle R \cong \angle M, \angle P \cong \angle S$;

$\dfrac{QR}{TM} = \dfrac{RP}{MS} = \dfrac{QP}{TS}$  **7.** $\angle A \cong \angle E, \angle B \cong \angle D$,

$\angle ACB \cong \angle ECD; \dfrac{AB}{DE} = \dfrac{BC}{DC} = \dfrac{AC}{EC}$

**8.** $\angle N \cong \angle N, \angle L \cong \angle NMO, \angle P \cong \angle NOM$,

$\dfrac{LP}{MO} = \dfrac{LN}{MN} = \dfrac{PN}{ON}$  **9.** *GHI*  **10.** *GI, HI, GH*

**11.** 12, $x$  **12.** 12, $y$  **13.** $13\frac{1}{3}, 10\frac{2}{3}$

## Practice B

**1.** $\angle N \cong \angle X, \angle T \cong \angle Y, \angle M \cong \angle B$;

$\dfrac{MN}{BX} = \dfrac{NT}{XY} = \dfrac{MT}{BY}$  **2.** $\angle BAD \cong \angle CAE$;

$\angle ABD \cong \angle ACE; \angle ADB \cong \angle AEC$;

$\dfrac{AB}{AC} = \dfrac{BD}{CE} = \dfrac{AD}{AE}$  **3.** $\angle G \cong \angle L, \angle H \cong \angle LIK$,

$\angle GJH \cong \angle K; \dfrac{GH}{LI} = \dfrac{HJ}{IK} = \dfrac{GJ}{LK}$  **4.** *AGN*

**5.** *AG, GN, NA*  **6.** 16, $x$  **7.** 16, $y$  **8.** 15, 18

**9.** yes; $\triangle NMO \sim \triangle QRP$  **10.** not enough information to determine  **11.** yes; $\triangle DFG \sim \triangle RST$

**12.** yes; $\triangle ABE \sim \triangle DBC$

**13.** yes; $\triangle XYW \sim \triangle ZYV$

**14.** yes; $\triangle SRT \sim \triangle SQU$

**15.** *Sample answer:*

| Statements | Reasons |
|---|---|
| **1.** $\overline{DE}$ is midsegment of $\triangle ABC$. | **1.** Given |
| **2.** $\overline{DE} \parallel \overline{AC}$ | **2.** Midsegment Thm. |
| **3.** $\angle CAB \cong \angle EDB$ | **3.** Corresp. $\angle$s Post. |
| **4.** $\angle B \cong \angle B$ | **4.** Reflexive Prop. of Congruence |
| **5.** $\triangle ABC \sim \triangle DBE$ | **5.** AA ~ Postulate |

## Practice C

**1.** $\angle A \cong \angle Q, \angle B \cong \angle R, \angle C \cong \angle T$;
$\triangle ABC \sim \triangle QRT$  **2.** $\angle S \cong \angle UTV$,

$\angle W \cong \angle UVT, \angle TUV \cong \angle SUW$;
$\triangle SUW \sim \triangle TUV$  **3.** $\angle R \cong \angle M, \angle L \cong \angle MON$,

$\angle MNO \cong \angle RTL, \triangle RLT \sim \triangle MON$  **4.** not
enough information  **5.** yes; $\triangle LMN \sim \triangle HGD$

**6.** yes; $\triangle XTR \sim \triangle KAJ$  **7.** yes; $\triangle QNM \sim \triangle PNO$

**8.** yes; $\triangle ABC \sim \triangle EDC$  **9.** yes;
$\triangle RSV \sim \triangle RTU$  **10.** $x = 5, y = \frac{1}{2}\sqrt{149}$

**11.** $x = \frac{10}{3}, y = \frac{14}{3}$

**12.**

| Statements | Reasons |
|---|---|
| **1.** $\angle ABC$ is a right $\triangle$. | **1.** Given |
| **2.** $\overline{AD}$ is an altitude. | **2.** Given |
| **3.** $\overline{AD} \perp \overline{BC}$ | **3.** Def. of altitude |
| **4.** $\angle CDA$ is a right $\angle$. | **4.** $\perp$ lines intersect to form right $\angle$s. |
| **5.** $\angle CAB \cong \angle CDA$ | **5.** All right $\angle$s are $\cong$. |
| **6.** $\angle C \cong \angle C$ | **6.** Reflexive Prop. of Congruence |
| **7.** $\triangle ABC \sim \triangle DAC$ | **7.** AA ~ Postulate |

## Reteaching with Additional Practice

**1.** $\angle A$ and $\angle P, \angle C$ and $\angle N, \angle B$ and $\angle M$;

$\dfrac{AB}{PM} = \dfrac{BC}{MN} = \dfrac{CA}{NP}$; 30  **2.** $\angle Y$ and $\angle Z, \angle T$ and

$\angle W; \dfrac{XY}{XZ} = \dfrac{YT}{ZW} = \dfrac{TX}{WX}$; 4.25

# Lesson 8.4 *continued*

**3.** $\angle K$ and $\angle M$,

$\angle L$ and $\angle J$; $\dfrac{KL}{MJ} = \dfrac{LJ}{JL} = \dfrac{JK}{LM}$; 8.2

**4.** $ABD \sim BCE$   **5.** no; the angles in $\triangle PQN$ are not congruent to the angles in $\triangle MPR$.

**6.** $XYT \sim XWZ$

## Interdisciplinary Application

**1.**

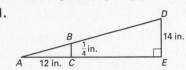

**2.** $\overline{BC}$   **3.** $\overline{AC}$   **4.** 55 feet   **5.** 287 feet

## Challenge: Skills and Applications

**1. a.** $\triangle VWX \sim \triangle ZYX$   **b.** *Sample answer:* Since $\overline{VW} \parallel \overline{YZ}$, the Alternate Interior Angles Theorem gives $\angle V \cong \angle Z$ and $\angle W \cong \angle Y$. So, by the AA Similarity Postulate, $\triangle VWX \sim \triangle ZYX$.

**2. a.** $\triangle AEH$, $\triangle DGH$, $\triangle CGF$   **b.** *Sample answer:* Prove $\triangle BEF \sim \triangle AEH$: Since $ABCD$ is a parallelogram, $\overline{BC} \parallel \overline{AD}$. So, by the Corresponding Angles Postulate, $\angle EBF \cong \angle A$. Also, by the Reflexive Property of Congruence, $\angle E \cong \angle E$. So by the AA Similarity Postulate, $\triangle BEF \sim \triangle AEH$. Prove $\triangle BEF \sim \triangle DGH$: Since $ABCD$ is a parallelogram, $\overline{BC} \parallel \overline{AD}$ and $\overline{AB} \parallel \overline{DC}$. So by the Corresponding Angles Postulate, $\angle E \cong \angle DGH$ and $\angle BFE \cong \angle H$. So, by the AA Similarity Postulate, $\triangle BEF \sim \triangle DGH$. Prove $\triangle BEF \sim \triangle CGF$: Since $ABCD$ is a parallelogram, $\overline{AB} \parallel \overline{DC}$. So, by the Alternate Interior Angles Theorem, $\angle E \cong \angle CGF$. Also, by the Vertical Angles Theorem, $\angle BFE \cong \angle CFG$. So, by the AA Similarity Postulate, $\triangle BEF \sim \triangle CGF$.

**3.** $(6, 4), (6, -4)$   **4.** $(0, 9), (0, -9)$

**5.** $\left(\dfrac{54}{13}, \dfrac{36}{13}\right), \left(\dfrac{54}{13}, -\dfrac{36}{13}\right)$   **6.** $(0, 4), (0, -4)$

**7.** $(6, 9), (6, -9)$   **8.** $\left(\dfrac{24}{13}, \dfrac{36}{13}\right), \left(\dfrac{54}{13}, -\dfrac{36}{13}\right)$

**9.** false; *Sample answer:* Let $ABCD$ be a square and let $EFGH$ be a nonsquare rectangle.

**10.** 450,000 km

# Lesson 8.5

## Warm-Up Exercises

**1.** $\triangle LMP \sim \triangle MQP$

**2.** by the AA Similarity Postulate   **3.** 7.5

## Daily Homework Quiz

**1.** $XYZ$   **2.** $\dfrac{5}{6}$   **3.** 10   **4.** 9   **5.** 100°

## Lesson Opener

Allow 15 minutes.

**1–4.** Check work.

## Technology Activity

**1.** yes   **2.** SSS Similarity Theorem

**3.** corresponding angles

**4.** AA Similarity Postulate

## Practice A

**1.** SAS Similarity Theorem; $\triangle FDM \sim \triangle AVQ$

**2.** SSS Similarity Theorem; $\triangle CEA \sim \triangle CDB$

**3.** AA Similarity Postulate; $\triangle AMT \sim \triangle HUT$

**4.** $\triangle DEF \sim \triangle GHI$; 2:1   **5.** $\triangle MNO \sim \triangle PQR$; 3:2

**6.** no   **7.** yes; $\triangle TCM \sim \triangle ASR$; SSS Similarity Theorem   **8.** True; since the right $\angle$s are congruent, use AA Similarity Postulate.   **9.** True; corresponding sides would be proportional so use SSS Similarity Theorem.   **10.** True; corresponding angles would be congruent so use AA Similarity Postulate.   **11.** false   **12.** True; ccorresponding legs would be proportional so use SAS Similarity Theorem.

## Practice B

**1.** SSS Similarity Theorem; $\triangle ABC \sim \triangle XYZ$

**2.** SAS Similarity Theorem; $\triangle QNM \sim \triangle ONP$

**3.** AA Similarity Postulate; $\triangle DEH \sim \triangle DFG$

**4.** no

**5.** yes; $\triangle RDM \sim \triangle SXT$; AA Similarity Postulate

**6.**        **7.**

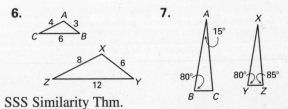

SSS Similarity Thm.

AA Similarity Postulate

# Lesson 8.5 *continued*

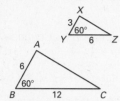

**8.**

SAS Similarity Theorem

**9.** △CED  **10.** 44°
**11.** 68°  **12.** 20
**13.** 5:2  **14.** 72 ft
**15.** 36 ft

## Practice C

**1.** yes; △ACB~△DCE; AA Similarity Postulate

**2.** no  **3.** yes; △DMP~△LMN; SAS Similarity Theorem

**4.**                                    **5.**

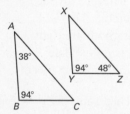                     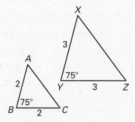

AA Similarity Postulate

                                         SAS Similarity Theorem

**6.**                                   **7.** 45°  **8.** 85°

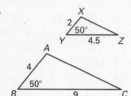                     **9.** 10  **10.** $10\sqrt{2}$

                                         **11.** $10 + \sqrt{69}$

SAS Similarity Theorem

**12.** △ABD~△GFD, △CBD~△EFD, △ACD~△GED

**13.**

| Statements | Reasons |
|---|---|
| 1. △ABC is equilateral. | 1. Given |
| 2. AB = BC = AC | 2. Def. of equilateral △ |
| 3. $\overline{DE}$, $\overline{DF}$, and $\overline{EF}$ are midsegments. | 3. Given |
| 4. $DE = \frac{1}{2}BC$, $EF = \frac{1}{2}AC$, $DF = \frac{1}{2}AB$ | 4. Midsegment Thm. |
| 5. △ABC~△FED | 5. SSS Similarity Thm. |

**14.**

| Statements | Reasons |
|---|---|
| 1. ABCD is a trapezoid. | 1. Given |
| 2. $\overline{AD}$ and $\overline{BC}$ are bases. | 2. Given |
| 3. $\overline{AD} \parallel \overline{BC}$ | 3. Def. of base of trapezoid |
| 4. ∠EDA ≅ ∠ECB | 4. Corresponding △ Postulate |
| 5. ∠EAD ≅ ∠EBC | 5. Corresponding △ Postulate |
| 6. △EAD ≅ △EBC | 6. AA Similarity Postulate |

## Reteaching with Additional Practice

**1.** ABC ~ EFD  **2.** MNP ~ RQS  **3.** The two pairs of corresponding sides which form the right angles are proportional, so the triangles are similar by the SAS Similarity Theorem.  **4.** Because $\overline{PQ}$ and $\overline{RN}$ are parallel, corresponding angles are congruent. Now, the two pairs of corresponding sides which include the two corresponding angles Q and NRM are proportional, so the triangles are similar by the SAS Similarity Theorem.  **5.** The two triangles have an equal angle at point S because the angles are vertical angles. The two sides forming these angles are proportional, so the triangles are similar by the SAS Similarity Theorem.

## Cooperative Learning Activity

**1.** *Sample answer:* The objects are the same height using both methods.

## Real-Life Application

**1.**

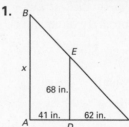

**2.** yes; AA Similarity Postulate

**3.** 113 in.

**4.** no

**5.** 109 in.

# Lesson 8.5 *continued*

## Math and History Applications

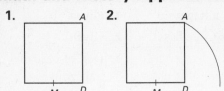

1. 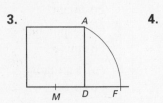 2.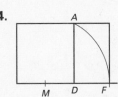

3. 4.

5. Answers will vary.

## Challenge: Skills and Applications

1. a. *Sample answer:*

| Statements | Reasons |
|---|---|
| 1. $\overline{BC} \parallel \overline{DE}$ | 1. Given |
| 2. $\angle ABC \cong \angle ADE$ | 2. Corres. $\angle$s Postulate |
| 3. $\angle A \cong \angle A$ | 3. Reflexive Property of Congruence |
| 4. $\triangle ABC \sim \triangle ADE$ | 4. AA Similarity Postulate |

b. *Sample answer:*

| Statements | Reasons |
|---|---|
| 1. $\overline{BC} \parallel \overline{DE}$ | 1. Given |
| 2. $\angle BCF \cong \angle EDF$, $\angle CBF \cong \angle DEF$ | 2. Alternate Interior Angles Theorem |
| 3. $\triangle BCF \sim \triangle EDF$ | 3. AA Similarity Postulate |

c. *Sample answer:* Since $\triangle ABC \sim \triangle ADE$,

$\dfrac{AC}{AE} = \dfrac{BC}{DE}$. Since $\triangle BCF \sim \triangle EDF$, $\dfrac{BF}{FE} = \dfrac{BC}{DE}$.

Therefore, by the transitive property of equality,

$\dfrac{AC}{AE} = \dfrac{BF}{FE}$.   2. *Sample answer:* Since

$\triangle GHI \sim \triangle KLM$, $\angle G \cong \angle K$ and

$\angle GHI \cong \angle KLM$. But $\overrightarrow{HJ}$ bisects

$\angle GHI$ and $\overrightarrow{LN}$ bisects $\angle KLM$, so

$\angle 1 \cong \angle 3$. Therefore, by the AA Similarity

Postulate, $\triangle GHJ \sim \triangle KLN$. So, $\dfrac{HJ}{LN} = \dfrac{GH}{KL}$.

3. *Sample answer:* Since $\overline{OG}$ is a median of $\triangle PGR$ and $\overline{SU}$ is a median of $\triangle TUV$, we know that $O$ is the midpoint of $\overline{PR}$ and $S$ is the midpoint of $\overline{TV}$. Therefore, $PR = 2 \cdot PO$ and $TV = 2 \cdot TS$.

So, $\dfrac{PR}{TV} = \dfrac{2 \cdot PO}{2 \cdot TS} = \dfrac{PO}{TS}$. Since $\triangle PGR \sim \triangle TUV$,

$\dfrac{PG}{TU} = \dfrac{PR}{TV}$; by the transitive property of equality,

$\dfrac{PG}{TU} = \dfrac{PO}{TS}$. We also know that $\angle P \cong \angle T$, so by

the SAS Similarity Theorem, $\triangle OPG \sim \triangle STU$.

This gives $\dfrac{OG}{SU} = \dfrac{PO}{TS}$. By the transitive property

of equality, $\dfrac{OG}{SU} = \dfrac{PR}{TV}$.   4. *Sample answer:* Since

$\triangle WXY$ is a right triangle, $\angle W$ and $\angle Y$ are complimentary. Since $\triangle WXZ$ is a right triangle, $\angle W$ and $\angle 1$ are complementary. Therefore, by the Congruent Complements Theorem, $\angle 1 \cong \angle Y$. Also, $\angle XZW \cong \angle YZX$, because both are right angles. So by the AA Congruence Postulate, $\triangle WXZ \sim \triangle XYZ$.   5. a. *Sample answer:* Since

$\triangle WXZ \sim \triangle XYZ$, $\dfrac{WZ}{XZ} = \dfrac{XZ}{YZ}$. Therefore, $(XZ)^2 =$

$(WZ)(ZY)$.   b. *Sample answer:* The AA

Congruence Postulate can be used to show that

$\triangle WZX \cong \triangle WXY$. Therefore, $\dfrac{WZ}{WX} = \dfrac{WX}{WY}$. So,

$(WX)^2 = (WY)(WZ)$.   c. *Sample answer:* The AA Congruence Postulate can be used to show that

$\triangle XZY \cong \triangle WXY$. Therefore, $\dfrac{ZY}{XY} = \dfrac{XY}{WY}$. So,

$(XY)^2 = (WY)(ZY)$.   d. *Sample answer:*
$(WX)^2 + (XY)^2 = (WY)(WZ) + (WY)(ZY)$
$\qquad\qquad = (WY)(WZ + ZY)$
$\qquad\qquad = (WY)(WY) = (WY)^2$

## Quiz 2

1. yes; $m\angle DCG = 95°$, $m\angle D = 50°$, $m\angle A = m\angle G = 35°$

2. yes; $m\angle R = m\angle S = 90°$, $m\angle T = 25°$, $m\angle Q = m\angle TRS = 65°$   3. no; $m\angle X = 90°$, $m\angle Y = 45°$, $m\angle B = m\angle C = 44°$   4. yes

5. no   6. yes   7. 120 yds

## *Lesson 8.6*

## Warm-Up Exercises

1. $\frac{4}{3}$   2. 49.5   3. 1.4   4. 2.5

## Lesson 8.6 *continued*

### Daily Homework Quiz

1. yes; $\triangle FGE \sim \triangle JHI$; SSS Similarity Thm.
2. yes; $\triangle ABC \sim \triangle XYZ$; SAS Similarity Thm.
3. 240 ft

### Lesson Opener

Allow 5 minutes.

1. C, E  2. B, F  3. A, D
4. CELEBRATE! HAVE FUN! YOU ARE DONE!

### Practice A

1. $FG$  2. $CE$  3. $CE$  4. $AF$  5. $AF$
6. $AC$  7. True; $\triangle$ Proportionality Thm.
8. False; it should be $\dfrac{AC}{AE} = \dfrac{BC}{DE}$.
9. False; it should be $\dfrac{EA}{CA} = \dfrac{ED}{CB}$.
10. True; $\triangle$ Proportionality Thm.  11. Yes; SAS Similarity Thm. so $\angle XZY \cong \angle XWV$ so corresponding $\angle$s are $\cong$ and lines are parallel.
12. Yes; Converse of $\triangle$ Proportionality Thm.
13. Yes; $\angle XZY \cong \angle XWV$, corresponding $\angle$s are $\cong$ and lines are parallel.  14. no  15. A  16. B
17. D  18. C  19. 4.5  20. 7.5  21. 6

### Practice B

1. $JK$  2. $NO$  3. $MK$  4. $PM$  5. $MP$
6. $JN$  7. No; $\frac{7}{2} \neq \frac{8}{3}$.  8. Yes; $\overline{DE}$ divides two sides of $\triangle ABC$ proportionally.  9. Yes; $\overline{DE}$ divides two sides of $\triangle ABC$ proportionally.
10. 4.5  11. $4\frac{2}{3}$  12. $6\frac{2}{3}$  13. 15  14. $9\frac{1}{3}$
15. $7\frac{5}{7}$  16. $3\frac{1}{3}$
17.

| Statements | Reasons |
|---|---|
| 1. $\overline{GB} \parallel \overline{FC} \parallel \overline{ED}$ | 1. Given |
| 2. $\angle ABG \cong \angle ADE$ | 2. Corresponding $\angle$s Post. |
| 3. $\angle GAB \cong \angle EAD$ | 3. Reflexive Prop. of $\cong$ |
| 4. $\triangle ABG \sim \triangle ADE$ | 4. AA Similarity Post. |

### Practice C

1. $AG$  2. $DE$  3. $EG$  4. $BG$  5. $CG$
6. $DC$  7. 10  8. 20  9. 10  10. $2\frac{2}{3}$

11. 7.5  12. 11.25  13. 10  14. $11\frac{1}{5}$  15. 12.5
16. 4.5
17.

| Statements | Reasons |
|---|---|
| 1. $\overline{WZ}$ bisects $\angle XZY$. | 1. Given |
| 2. $\dfrac{XW}{XZ} = \dfrac{WY}{ZY}$ | 2. Bisector of an $\angle$ divides the opposite side into segments whose lengths are proportional to the lengths of the other two sides. |
| 3. $XW \cdot ZY = XZ \cdot WY$ | 3. Cross products prop. |
| 4. $XW = WY$ | 4. Given |
| 5. $ZY = XZ$ | 5. Division prop. of equality |

### Reteaching with Additional Practice

1. 3.5  2. $\frac{200}{9}$  3. $x = 8.9, y = 14.1$
4. $a = 6.5$  5. $x = 2.7$  6. $x = 7.8, y = 7.2$

### Interdisciplinary Application

1. $\dfrac{10}{11} = \dfrac{x}{12}$; 10.9 millimeters
2. $\dfrac{y}{11} = \dfrac{9}{12}$; 8.25 millimeters
3. $\dfrac{6}{11} = \dfrac{z}{12}$; 6.5 millimeters

### Challenge: Skills and Applications

1. a. $\dfrac{AD}{BD} = \dfrac{AC}{BC}$  b. *Sample answer:* Since parallel lines divide transversals proportionally, $\dfrac{AE}{BE} = \dfrac{AC}{FC}$.  c. *Sample answer:* Since $\overleftrightarrow{BF} \parallel \overleftrightarrow{CE}$, $\angle 2 \cong \angle CFB$ by the Corresponding Angles Postulate and $\angle 1 \cong \angle CBF$ by the Alternate Interior Angles Theorem. But $\angle 1 \cong \angle 2$. Therefore, by the transitive property of congruence, $\angle CFB \cong \angle CBF$. Therefore, $\triangle FBC$ is an isosceles triangle (with $\overline{CF} \cong \overline{CB}$) by the Converse of the Base Angles Theorem.  d. Based on part (c), $FC = BC$. Substituting into the

# Lesson 8.6 *continued*

proportion from part (b), $\dfrac{AE}{BE} = \dfrac{AC}{FC}$, gives

$\dfrac{AE}{BE} = \dfrac{AC}{BC}$. But, from part (a), $\dfrac{AD}{BD} = \dfrac{AC}{BC}$; so

$\dfrac{AD}{BD} = \dfrac{AE}{BE}$ by the transitive property of equality.

**2.** *Sample answer:* If $AC = BC$, then the bisector of an exterior angle at $C$ is parallel to $\overleftrightarrow{AB}$, so point $E$ as described in Exercise 1 cannot exist. However, the *theorem* is still true! (A statement "$p$ implies $q$" is true whenever $p$ is false.)

**3.** $BC = 12, BE = 72$  **4.** $AC = 40, AD = 30$

**5.** $AD = 6, BE = 20$  **6.** $AC = 12, BE = 55$

**7.** $BC = 6, AE = 20$  **8.** $x = 3$  **9.** $x = 4$

**10.** $\sqrt{73}$  **11.** $m\angle ABC = 80°, m\angle BCE = 60°$

# Lesson 8.7

## Warm-Up Exercises

**1.** $\sqrt{13}$  **2.** $\sqrt{26}$  **3.** A figure is rotated around a fixed point.  **4.** A figure is reflected over a line.  **5.** A figure is shifted up or down and/or right or left.

## Daily Homework Quiz

**1.** 8  **2.** 6.3  **3.** no  **4.** $\dfrac{TV}{VS} = \dfrac{RT}{RS}$

## Lesson Opener

Allow 15 minutes.

**1.** They are equal to the scale factor.

**2.** reduced  **3.** enlarged

**4.** *Sample answer:*

$\triangle AED, \triangle EBF,$
$\triangle DFC$

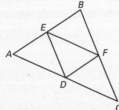

**5.** *Sample answer:*

$\triangle ADE, \triangle GBF,$
$\triangle HIC$

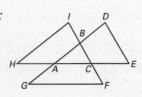

## Practice A

**1.** similar  **2.** smaller, a reduction  **3.** larger, an enlargement  **4.** The dilation has center $C$ and scale factor $\frac{1}{2}$.  **5.** The dilation has center $C$ and scale factor $\frac{15}{7}$.  **6.** The dilation has center $C$ and scale factor $\frac{5}{2}$.  **7.** $k = 3, x = 12, y = 9$

**8.** $k = 4, x = 8, y = 20, z = 8$

**9.** $k = 4, x = 4, y = 2, z = 4$

**10.** $M'(0, 8), N'(6, 8), L'(6, 0)$

**11.** $G'(1, 4), H'(3, 3), I'(2, 1)$

## Practice B

**1.** The dilation has center $C$ and scale factor $\frac{3}{2}$.

**2.** The dilation has center $C$ and scale factor $\frac{8}{3}$.

**3.** The dilation has center $C$ and scale factor $\frac{5}{9}$.

**4.** The dilation has center $C$ and scale factor $\frac{3}{1}$; $x = 15, y = 18$.  **5.** The dilation has center $C$ and scale factor $\frac{5}{12}$; $x = 2.5$.

**6.** $M'(0, 9), N'(6, 12), L'(12, 0)$

**7.** $G'\left(\frac{5}{3}, \frac{7}{3}\right), H'(3, 1), I'\left(1, \frac{1}{3}\right)$

**8.**

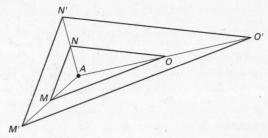

**9.**

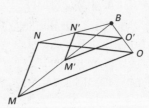

**10.** 6 in. by $7\frac{1}{2}$ in.

## Practice C

**1.** The dilation has center $C$ and scale factor $\frac{5}{2}$; $x = 20, y = 10$.  **2.** The dilation has center $C$ and scale factor $\frac{9}{4}$; $x = \frac{27}{4}, y = \frac{27}{2}$.

**3.** The dilation has center $C$ and scale factor $\frac{2}{3}$; $x = \frac{15}{2}, y = \frac{21}{2}, z = \frac{15}{2}$.  **4.** The dilation has center $C$ and scale factor $\frac{3}{5}$; $x = 11.04$.

**5.** $M'\left(-\frac{2}{3}, \frac{4}{3}\right), N'\left(\frac{4}{3}, \frac{8}{3}\right), L'(4, 0)$

**6.** $G'(-5, 5), H'(0, 15), I'(20, 12.5), J'(15, 2.5)$

# Lesson 8.7 *continued*

**7.** 36 in.

## Reteaching with Additional Practice

**1.** $k = \dfrac{3}{7}$, reduction   **2.** $k = 2$, enlargement

**3.** $k = \dfrac{7}{3}$, enlargement   **4.** $k = \dfrac{2}{5}$, reduction

**5.** $A'(0, 6)$, $B(6, 6)$, $C'(4.5, 3)$

**6.** $A'(-24, -9)$, $B'(-15, -12)$, $C'(-6, -9)$, $D'(-12, -3)$

**7.** $K'(0.5, 4)$, $L'(1, 2)$, $M'(-5, 0)$, $N'(-3, 0)$

**8.** $X'(-2.25, -1.5)$, $Y'(6, 3)$, $Z'(3, -3)$

## Real-Life Application

**1.** $\dfrac{4}{5} = 0.8$ or 80%   **2.** 3.2 inches   **3.** 7 dolls are inside the first one.   **4.** about 5 dolls

## Challenge: Skills and Applications

**1.**

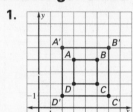

**2. a.** $(k(x - a) + a, k(y - b) + b)$

**b.** Answer should match Exercise 1.

**3.** $A'(-10, 25)$, $B'(10, 13)$, $C'(6, 1)$

**4.** $A'(-2, 13)$, $B'(13, 4)$, $C'(10, -5)$

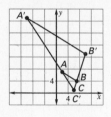

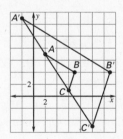

**5.** $A'(6, 1)$, $B'(16, -5)$, $C'(14, -11)$

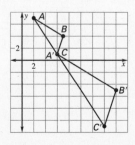

**6.** $D'(0, 9)$, $E'(8, 1)$, $F'(4, 1)$, $G'(0, 5)$

**7.** $D'(-1, 6)$, $E'(5, 0)$, $F'(2, 0)$, $G'(-1, 3)$

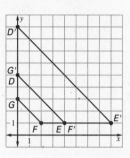

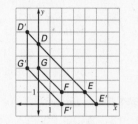

**8.** $D'(0, 7)$, $E'(2, 5)$, $F'(1, 5)$, $G'(0, 6)$

## Review and Assessment

### Review Games and Activities

**1.** 1   **2.** 1   **3.** 2   **4.** 3   **5.** $\dfrac{10}{3}$   **6.** 8
**7.** 13   **8.** 21   **9.** 34   **10.** 55°   **11.** 89

### Test A

**1.** 16   **2.** 75   **3.** 14   **4.** 7.2   **5.** 2   **6.** $4 + m$
**7.** 3.2   **8.** 4   **9.** 5.25   **10.** 16   **11.** $\dfrac{1}{2}$   **12.** 2
**13.** $s = 3.5, t = 7, u = 5$   **14.** $ABCD$: 15; $WXYZ$: 30   **15.** 1:2   **16.** no   **17.** yes; $\triangle GHI \sim \triangle JLK$; SSS Similarity   **18.** yes; $\triangle TUS \sim \triangle WUV$; AA Similarity   **19.** yes; $\triangle ABC \sim \triangle ADE$; AA Similarity   **20.** no
**21.** yes; $\triangle MNO \sim \triangle PQO$; SAS Similarity
**22.** $ER$   **23.** $AR$   **24.** $AR$   **25.** $AE$   **26.** $AR$
**27.** $DE$   **28.** 6.75   **29.** $8\frac{2}{11}$   **30.** 6

### Test B

**1.** 6   **2.** 10   **3.** 2   **4.** $\dfrac{1}{2}$   **5.** $\dfrac{s}{5}$   **6.** $\dfrac{n}{m}$
**7.** $4 + q$   **8.** 3   **9.** $\dfrac{15}{7}$   **10.** 13   **11.** $\dfrac{1}{2}$   **12.** 2
**13.** $s = 4.5, t = 16, u = 10$   **14.** $ABCD$: 21; $WXYZ$: 42   **15.** 1:2   **16.** no   **17.** yes; $\triangle GHI \sim \triangle KJL$; SAS Similarity   **18.** yes; $\triangle MON \sim \triangle QOP$; AA Similarity   **19.** yes; $\triangle RSV \sim \triangle UTV$; AA Similarity   **20.** no

# Review and Assessment *continued*

**21.** yes; △*JHI* ~ △*JFG*; AA Similarity

**22.** $\dfrac{GF}{FE}$  **23.** $\dfrac{GF}{FD}$  **24.** $\dfrac{AB}{AC}$  **25.** $\dfrac{ED}{GD}$  **26.** $\dfrac{DC}{AB}$

**27.** $\dfrac{AB}{CD}$  **28.** 7.2  **29.** 10  **30.** 15

## Test C

**1.** 18  **2.** 12  **3.** 12  **4.** 10  **5.** $\frac{32}{11}$ or $2\frac{10}{11}$

**6.** 33.75  **7.** $\dfrac{a}{2}$  **8.** $\dfrac{2m}{2n}$  **9.** $\dfrac{3+q}{q}$  **10.** 2

**11.** $7\frac{1}{2}$  **12.** $21\frac{1}{3}$  **13.** 26  **14.** $\frac{1}{3}$  **15.** 3

**16.** $r = 9, s = 13.5, t = 7, u = 5$

**17.** *ABCDE*: 23.5; *VWXYZ*: 70.5  **18.** 1:3

**19.** no  **20.** yes; △*GHI* ~ △*JLK*; AA Similarity

**21.** yes; △*MNO* ~ △*QPO*; AA Similarity

**22.** yes; △*RST* ~ △*RUV*; AA Similarity

**23.** yes; △*ABE* ~ △*CBD*; SSS Similarity

**24.** $\dfrac{GF}{FE}$  **25.** $\dfrac{GF}{FD}$  **26.** $\dfrac{AB}{AC}$  **27.** $\dfrac{ED}{GD}$  **28.** $\dfrac{CD}{AB}$

**29.** $\dfrac{AB}{CD}$  **30.** 6  **31.** $\dfrac{7}{3}$  **32.** 10

## SAT/ACT Chapter Test

**1.** A  **2.** C  **3.** C  **4.** D  **5.** B  **6.** B  **7.** E
**8.** A  **9.** D

## Alternative Assessment

**1.** Complete answers should include:

• Congruent angles ∠*A* ≅ ∠*M*, ∠*B* ≅ ∠*I*, ∠*C* ≅ ∠*J*, ∠*D* ≅ ∠*K*, ∠*E* ≅ ∠*L*.  • Statement of proportionality $\dfrac{AB}{MI} = \dfrac{BC}{IJ} = \dfrac{CD}{JK} = \dfrac{DE}{KL} = \dfrac{EA}{LM}$.
• Scale factor of 5 to 3.  • $w = 5.33, x = 3$, $y = 3$, and $z = 3.6$  • Perimeter of *ABCDE* is 15.8 and the perimeter of *MIJKL* is 26.33.

• The ratio of the perimeters is 3 to 5.

**2. a.** The scale factor of △*CAB* to △*CUP* is 1 to 2 and the scale factor of △*CAB* to △*COT* is 2 to 3.
**b.** 27 units  **c.** 18 units  **d.** 22.5 units
**e.** 30 units  **3. a.** *CN* = 5.7 units, *NB* = 6.3 units

**b.** If $\dfrac{CO}{OU} = \dfrac{CT}{TP}$, then $\overline{OT} \parallel \overline{UP}$.
$\dfrac{20.25}{6.75} = \dfrac{18}{6} \rightarrow 121.5 = 121.5$. Therefore, $\overline{OT}$ is parallel to $\overline{UP}$.  **4.** Answers may vary, the final statements should prove △*COT* ~ △*CUP*.

## Project: Self Similarity

**1.** Check drawings. I(0): 9-in. line segment

**2.** Check drawings. I(1): 9-in. line segment with 3-in. hole in the middle  **3.** Check drawings. Each row has the segments from the previous row with the middle third missing.

**4.**

| Iteration | Length of each segment | No. of segments in row | Total length shown in row |
|---|---|---|---|
| 0 | 1 | 1 | 1 |
| 1 | $\frac{1}{3}$ | 2 | $\frac{2}{3}$ |
| 2 | $\frac{1}{9}$ | 4 | $\frac{4}{9}$ |
| 3 | $\frac{1}{27}$ | 8 | $\frac{8}{27}$ |
| 4 | $\frac{1}{81}$ | 16 | $\frac{16}{81}$ |
| 5 | $\frac{1}{243}$ | 32 | $\frac{32}{243}$ |
| $n$ | $\left(\frac{1}{3}\right)^n$ | $2^n$ | $\left(\frac{2}{3}\right)^n$ |

**5.** *Sample answer:* Each new segment with a hole in it is similar to I(1) because it has a hole in it that is one third of the segment.

## Cumulative Review

**1.** $(2, 7)$  **2.** $\left(\frac{3}{2}, 2\right)$  **3.** Addition property of equality  **4.** Symmetric property of equality

**5.** $x = 60, y = 40$  **6.** $x = 12, y = 13$

**7.** 120°  **8.** 110°  **9.** $\overline{XZ}$  **10.** 12  **11.** 14

**12.** 19  **13.** $A'(-3, -1), B'(-1, -4)$

**14.** $A'(3, 1), B'(1, 4)$  **15.** 10.5  **16.** $\frac{1}{6}$  **17.** 12

**18.** $\frac{3}{2}$  **19.** $\frac{4}{3}$  **20.** 55°  **21.** 12

**22.** △*JKL* ≅ △*MNO*, SAS Similarity Theorem

**23.** △*PQR* ≅ △*STR*, AA Similarity Postulate

**24.** $\frac{10}{3}$  **25.** 18  **26.** $\frac{1}{2}$, reduction

**27.** $\frac{3}{2}$, enlargement